The Essential Guide to Building Your Argument

The Essential Guide to Building Your Argument

Dave Rush

Los Angeles | London | New Delhi
Singapore | Washington DC | Melbourne

SAGE Publications Ltd
1 Oliver's Yard
55 City Road
London EC1Y 1SP

SAGE Publications Inc.
2455 Teller Road
Thousand Oaks, California 91320

SAGE Publications India Pvt Ltd
Unit No 323-333, Third Floor, F-Block
International Trade Tower, Nehru Place
New Delhi 110 019

SAGE Publications Asia-Pacific Pte Ltd
3 Church Street
#10-04 Samsung Hub
Singapore 049483

Editor: Kate Keers
Assistant Editor: Sahar Jamfar
Production Editor: Neelu Sahu
Copyeditor: Peter Williams
Proofreader: Salia Nessa
Indexer: Melanie Gee
Marketing Manager: Maria Omena
Cover Design: Sheila Tong
Typeset by KnowledgeWorks Global Ltd

Library of Congress Control Number 2022945796

British Library Cataloguing in Publication data

A catalogue record for this book is available from the British Library

ISBN 978-1-5297-6792-6
ISBN 978-1-5297-6791-9 (pbk)

For Anwyn and Laurie, who are already better at arguing than I am, and for Robin, without whom nothing would be possible.

Contents

About the author

Dave Rush is head of Skills for Success, the academic skills and English language support centre at the University of Essex. He has a PhD in English Literature and has previously taught in Nigeria, Sri Lanka and at the University of Sussex. His research interests include postmodernism, cultural history and critical English for Academic Purposes (EAP).

Acknowledgements

I would like to thank all the colleagues I have worked with at the University of Essex in the Talent Development Centre and Skills for Success. Everything I know about academic skills and EAP teaching I have learnt from them. The exercise based on Lynne Pettinger's blog piece in Chapter 4 is adapted from lesson material originally developed by Liz Austin, who I would also like to thank for giving me a job at Essex in the first place.

Introduction: What is an argument?

This book is intended mainly for people who are either at, or are thinking about going to, university. While it is primarily aimed at those making the transition into studying, whether at undergraduate or postgraduate level, it should also be useful for anyone looking to improve their understanding of and ability to build arguments at any point in their academic journey.

The aim is not to give you a 'how-to' type of guide, a set of rules to memorise or a set of templates that you can follow. Instead, throughout this book, you will be asked to think about the questions that particular tasks and disciplines are born from, the questions you should be asking yourself in the various situations you will encounter during your studies, and why those questions are being asked. This will give you a framework for producing your own approach to understanding and building arguments at university, rather than providing you with a set of rules to learn and boxes to tick.

This is not a 'skills' book, then, in some typical senses, but is rather intended as a guide to understanding why arguments are so important to university study, and just as importantly to understanding both what that study looks like and how it and the university as a whole works. In other words, the aim is to give you the tools you need to decode and engage with the educational context you find yourself in, no matter where that might be, and to show the importance of arguments both to that academic context and, more broadly, to our understanding of the world as a whole.

Chapter 1 will use the question 'why are you here?' as the entry point into understanding what arguments are and why they are important at university. The question will be twofold, covering both why you might want to study at university and why you might want to read this book about arguments in order to support that study.

This debate will be used as an entry point into exploring what arguments are, and how they are used at university. It will also be used to explore how even the most simple-seeming questions can lead to a wide range of competing, yet equally valid responses, and the ways in which particular contexts shape those responses.

Chapter 2 will explore the different types of argument that you will encounter at university, both in terms of how arguments are constructed, but also the different academic genres in which they appear, and how different disciplinary contexts affect the nature of arguments.

The notion of a 'good argument' is a construct that differs from context to context and culture to culture. Different national and regional cultures have different criteria for what makes a good argument, and so do different educational cultures. In other words, both the subject you are studying and the place you are studying will affect what is considered to be a good argument, and to build a good one, you need to understand both.

Building on this consideration of how different disciplines and cultures approach the question of what makes a good argument, **Chapter 3** will explore how these approaches reflect different standpoints, and how different ideological and theoretical frameworks are themselves made up of, and the product of, arguments.

This will be used to demonstrate that there is no purely 'objective' position to argue from. While objectivity is the aim of academic study and can be achieved to an extent, a position is always taken within, and influenced by, a wider framework. Understanding those frameworks will greatly improve the ability to both understand and utilise arguments. Understanding how these frameworks have been and are constructed will also give you a much better understanding of what universities are, how they work and what your own place is within that.

Arguments appear at all stages and in all areas of your studies. While it is easy to think about learning a subject as simply about gaining knowledge and learning facts, as the first three chapters will show, this knowledge is only useful in terms of how it can be used and applied, and is also only produced in the first place in response to the constantly repeating process of question and argument that this use and application necessitates.

In other words, in order to be a successful student at university you need to think about arguments as not just something that you will produce in your essays, assignments or presentations, but as a constant feature of the academic context, something that surrounds you at all times, and which you need to always be alive to.

To explore this in more detail, the second section of the book will look at how arguments work and how you can understand the mechanics of them. It will do this by focusing on three areas.

Chapter 4 will look at how to pick out arguments in academic situations, both in terms of how to unpack complex texts but also in terms of how to read/listen for argument and not simply for content. Using example texts, we will explore what it means to be an 'active reader', and how being alive to the presence of arguments will enable you to read more effectively and more critically.

Chapter 5 will build on this exploration of active reading by exploring how we engage with arguments in order to test their validity and generate counterarguments. A key part of engaging with and producing arguments is the ability to test and validate propositions, whether they are your own or others. This section will look at how to do this and show that, while 'counterarguments' can be about explicit disagreements, they can also be about simply demonstrating the different ways that it is possible to look at a question or to improve upon or add to an argument rather than simply refuting it.

This chapter will explore how this works in writing and other contexts and will look at identifying and generating counterarguments. It will also demonstrate how the consideration of challenges and counterarguments is a key part of the process of constructing good arguments in an academic context.

Both chapters 4 and 5 will look at useful strategies and common 'trigger questions' that you can apply to any text or argument that you approach in order to ensure that you are fully understanding its argument, and the framework from which it has emerged.

Having explored how arguments appear and can be identified, and how counterarguments can be identified and generated, **Chapter 6** will look at how to structure arguments clearly, and how to incorporate counterarguments in order to strengthen your own position.

The process of structuring an argument will be examined from the simplest everyday context, to show how academic arguments are related to, and often born out of more prosaic questions and that structuring an argument is a skill that everyone already has, whether they realise it or not. Real university-level essay questions will then be used to demonstrate how this everyday skill maps onto the academic skill, and how each stage of structuring is in fact a response to a set of questions about what you want to achieve, rather than a step in a predetermined process. There is no magic recipe book of structures or formulae for what makes good writing, but understanding how different ways of constructing your arguments affects their impact is vital in enabling you to produce the most effective work you can for your studies.

The effectiveness of arguments in an academic context is not just about what they contain, but also about how they are communicated. While the ideas, data and evidence included and the claims made are of vital importance, the impact of those claims will be crucially affected by the means used to present them.

That means that, as a student, you need to think about not just what you want to say, but how you are going to say it, whether that be verbally or in writing. Universities are there to create and communicate knowledge and as a member of the academic community, good communication is thus a key part of what is expected of you, no matter what subject you study. It is also one of the key skills that you will take from university into the rest of your life, whatever path you follow.

Some of the language and structures that you will use as a student will be determined by the discipline that you are pursuing and the specific institution and country in which you are studying. No matter where or what you are studying, however, there are certain key questions that you can ask yourself and strategies that you can use to build the most effective and impactful arguments. These strategies will also help you to be an effective communicator in all aspects of your life.

Chapter 7 will explore the practical process of using language to produce a written argument, whatever the genre or discipline. To do this, it will focus on the key questions around what the purpose of academic writing is, who the audience is in an academic context, who the writer is and how the answers to these questions affect linguistic choices, i.e. the actual words you put down on paper and why.

While a lot of focus is often placed on how to *write* good arguments at university, it is also important to consider how to argue well verbally. **Chapter 8** will look at giving good presentations (at all levels), therefore, but will also look at how to have meaningful and constructive discussions and debates. This will be related to more everyday contexts to show how the ability to discuss and disagree is vital in helping to produce new knowledge, and will look at both how to use language to construct verbal arguments and the etiquette around doing so.

Chapter 9, meanwhile, will look at how to make arguments your own, whether through the written or the spoken word. University mark schemes, at both undergraduate and postgraduate level, often talk about 'originality' or 'criticality'. This can be intimidating, but it need not be, and how to satisfy this criterion is key in understanding both how to build good arguments and how to do well at university.

However, despite how it is often phrased, being asked to demonstrate 'originality' is not about coming up with something entirely new, it is about showing that what is in your argument is not just things that other people have said. As such, this section will look at how to evidence arguments effectively, but also to ensure that that evidence is there to support, not *be*, your argument.

Successful university study is not about simply learning knowledge and facts and being able to repeat them back later. It is about being able to take that knowledge and use it to answer specific questions and generate new questions, and therefore new knowledge. Academic disciplines are an ongoing conversation between the existing and the new, and in order to help you to take part in that conversation, this book will look at how you can learn to speak, and argue, in the right language. Doing so will inevitably involve you working it out for yourself to some extent, and as with any language, you learn best from using it. But what you learn here will hopefully give you some strategies, some questions and some ways of thinking that allow you to engage in your own academic conversations – and arguments – with confidence.

1

Why are you here?

The word 'argument' does not mean the same thing in an academic context as it does in everyday speech. Rather than meaning a disagreement or debate, an 'argument' for university purposes is a position taken in response to a particular question or issue, supported by propositions or reasons why that position is convincing or true.

The term 'argument' refers to both your overall answer/position, therefore, and all the steps that you go through to reach that position. That is, when lecturers talk about 'your argument' they mean both your final destination and how you got there. Every point you make needs to be supported by evidence and ordered logically to reach a clear conclusion.

Arguments are all about questions, then, and how to answer them. In order to look at how arguments are built, it makes sense to start with a question, in this case – why are you here?

This question has two parts:

1. Why are you at university, or wanting to go to university?
2. Why are you reading a book about the importance of arguments at university?

In order to make answering these easier, let's simplify them a little, or perhaps more accurately, ask similar questions that are easier to respond to.

1. Why do you think going to university is a good idea?
2. What do you want to get out of reading this book?

TASK 1.1

Take some notes on your answer to these two questions. These notes do not have to be in complete sentences or connected. Try and write around 100 words for question 1, and to think of at least three goals for question 2.

There is no one 'right' answer to either of these questions, but it is vital that we consider both, as in order to have found yourself at this point, reading this book, it is likely that you are either at university or are considering it strongly, and that you have a sense that argumentation is in some way important to being successful in the studies that you will undertake there.

Why is going to university a good idea?

We will return to question 2 shortly, but for now let us focus on question 1.

Before you go any further, condense the ideas you have just been thinking about and taking notes on into one sentence of no more than 20–30 words.

Why do I think that going to university is a good idea?

____

____

____

____

We will return to this later in the chapter. For now, just be aware that what we have created here is the beginnings of an argument – your answer to a particular question (your sentence), and the set of reasons which support that answer (your notes).

Common responses to the question of why going to university is a good idea include the love of a subject, an interest in learning about something in particular, embarking on a particular career or just getting a job more generally, with a university education seen as likely to lead to better opportunities and salaries.

This framing of the response is common in the UK, with the purpose of university posed as an opposition between interest in (or love of) a subject on one side, and pursuit of a particular (or at least well paid) career on the other. In other words, it is money vs knowledge, with most people able to situate themselves somewhere on the spectrum between them, even if they are not fully one or the other.

Take a look at your notes on question 1 and see if you think your response can be mapped on to this **binary**. It is unlikely that you would fully identify with either end, and more likely that you would place yourself somewhere along the line between the two poles. Place a cross where you think you would be, and think about whether your proximity to one end

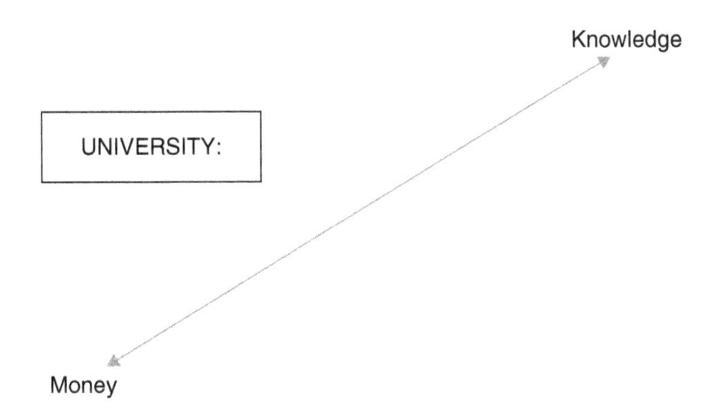

Figure 1.1 A common way of framing the debate about the purpose of university – where are you on the line?

or the other was fully reflected in the one-sentence answer you wrote a moment ago.

> A **binary**, or **binary opposition**, is a pair of opposing terms used to define the two poles of an argument. They are a common feature of thought across all disciplines. Examples could include capitalist/socialist, realist/abstract, quantitative/qualitative or right-wing/left-wing.

Like all binaries, this opposition represents an oversimplification of something much more complicated, but such simplifications can be a useful place to start. Other traditional reasons for going to university, in the UK and other developed nations, relate to achieving independence and 'finding yourself', having fun, socialising and building networks, gaining access to opportunities such as studying abroad, or perhaps even playing a particular sport to a higher level.

Your notes may have included some of these reasons, or a variation on them. My answer when I started university as an undergraduate at the turn of this millennium would have been a mixture of a love of my subject (literature) and a desire to become independent in a version of the world with less consequences than the 'real world' (by which I meant everywhere outside of the university).

But this is just one way of answering the question, and it is one that is born out of a very specific context. I am a white, middle-class, middle-aged, cis-gendered, straight man from one of the richest nations on Earth, which has a history of recent colonial domination. I speak what is arguably the dominant global language of academia and went to university in the UK. I now teach in a UK university.

All of this not only shapes my answer to the question, it also fundamentally shapes the very possibilities which I think are available in answering that question, some of which we have discussed above. The binary of money vs knowledge, for example, has a very specific historical and political basis, and is speaking to a particular, very passionate debate about the purpose of university (see, for example Collini, 2012). That debate covers fundamental questions about the purpose of university, and whether the joyful pursuit of knowledge and the utility and economic value of that knowledge are in conflict, or even fundamentally irreconcilable. But while anyone with an investment in that debate would seek to cast it as universal, its relevance in fact varies greatly depending on the individual student, the specific university system they study within and their reasons for undertaking that study.

I have no idea who 'you' reading this are, but you will have a very different set of possibilities to draw on in order to decide why university is a good idea. Some of those – what we might think of as 'facts', for example – will be the same for both of us. But many will be different, and even those which are the same will be radically different because of our different contexts, and changed by their relationship to the other elements of those contexts.

What this tells us, in terms of our consideration of arguments in general is that when we think about both understanding and building arguments, we are not just dealing with *facts or knowledge*, we are dealing with *contexts and approaches* – and often, it is very hard to tell the two apart.

What is university for?

A moment ago I asked you to write a one-sentence answer to the question: Why do you think university is a good idea? In a university context, this sentence would sometimes be referred to as a *thesis statement* – that is a clear statement of your central argument or point. The notes that you based that sentence on, consciously or not, are likely to contain a number of reasons why you think that answer is true. These could be referred to as the *premises* or *propositions* that support your thesis or answer. That answer and those propositions together form an *argument*, and it is likely to be one that feels quite straightforward to you. However, to see why it is perhaps more complex than you think, let us consider some of the different factors that have shaped both your overall position on university and your reasons for holding that position.

Your view of the university as an institution, for example, will be shaped by factors including your country of origin, whether you are intending to study in your own country or abroad, your class background, ethnicity, age and gender, whether anyone else in your family has been to university before and your own educational history, among many others. Universities are also often posited not simply as institutions, but as a means of moving from one situation to another – of, for example, escaping a position in society shaped by class, ethnicity or gender (for example) and wealth (or lack of it), or of a position in the world shaped by living in a less developed and not a developed country. Your view on this will also shape why you think going to university might be beneficial.

Your purpose in attending university, and the specific institution that you choose will be affected by all of these factors. Arguably, none of them change the 'facts' of what university is, but already we can see that within the easily discussed generalisation of the category named 'university' there is a wide range of specific variations – a university in London is different to a university in Lagos, to a university in LA or Lebanon, and so on.

Equally, your view of what university is *for* – what its purpose is – will vary greatly. In a purely practical sense we could map the purposes of the university on to the possible reasons for attending one. University is about the study of particular disciplines, and the production of new knowledge and ideas in those fields; it is about the preservation and production of knowledge, and the production of new fields and *types* of knowledge as well. It is also about equipping individuals with the ability to perform certain jobs, be that as an engineer, a nurse or an English literature teacher, and in this sense it performs both an individual and a social/economic role. That is, it does something for the individual student, but it also serves a function for the wider society, the nation and its economy – or the world and its economy, depending on your perspective.

The university also plays a social role, and that in itself can mean a number of different things. This could refer to giving people the opportunity to 'find themselves', to explore who they are and who they want to be. It could be about the creation of networks – both in the sense that it creates academic and research communities, but also that it provides personal connections that will make opportunities available in later life. It could also be about the notion that university fundamentally shapes the worldview of individuals, leading to either the creation of free-thinking individuals, or identical neoliberal units ready to be fed into, and consumed by, the globalised economy, depending on your point of view.

The university is often cast as a force for good, or at least as having a generally positive influence on both individuals and society. But consider the following passage from Ngugi wa Thiong'o's *Decolonising the Mind* (1986). What does his argument suggest about the role of university in Kenya under British colonial rule?

> ... the colonial system of education in addition to its apartheid racial demarcation had the structure of a pyramid: a broad primary base, a narrowing secondary middle, and an even narrower university apex. Selections from primary into secondary were through an examination, in my time called Kenya African Preliminary Examination, in which one had to pass six subjects ranging from Maths to Nature Study and Kiswahil. All the papers were written in English. Nobody could pass the exam who failed the English language paper no matter how brilliantly he had done in the other subjects. I remember one boy in my class of 1954 who had distinctions in all subjects except English, which he had failed. He was made to fail the entire exam. He went on to become a turn boy in a bus company. I who had only passes but a credit in English got a place at the Allance High School, one of the most elitist institutions for Africans in colonial Kenya. The requirements for a place at the University, Makerere University College, were broadly the same: nobody could go on to wear the undergraduate red gown, no matter how brilliantly they had performed in all the other subjects unless they had a credit – not even a simple pass! – in English. Thus the most coveted place in the pyramid and in the system was only available to the holder of an English language credit card. English was the official vehicle and the magic formula to colonial elitedom. (Ngugi wa Thiong'o, 1986, p12)

Here, far from being a positive force, the university is a part of a colonial system whereby the imposition of a language – English – was used to 'control ... the mental universe of the colonised', and 'how people perceived themselves and their relationship to the world' (p16). That colonial system might no longer exist in its historical form, but it, and others like it, continue to have a huge impact on the world. Colonialism, and the role that education played within it, represent an important part of the history of the university, and illustrate that the impact that university has on individual students, and the role it plays in their lives, goes far beyond what they might perceive as their individual reasons for attending such an institution. Your relationship to colonial structures is likely to be different to Ngugi's, but they and their effects remain a very real part of the world, and will have had a fundamental effect on your perception of yourself, your world and also of university.

Hierarchies of questions

Note how the question we started with above – why do you think going to university is a good idea? – has quite quickly led to the need to answer a number of other questions, some of which are very complex. These questions include, but are not limited to, what is a university, what is it for and what do you want to get out of it?

This shows how the argument we build in response to a question is perhaps not always directly a response to the original question – because that question is broader, or difficult to define, or itself contains multiple questions. The specific answer we choose is often actually a response to one aspect of the many hypothetical ways of approaching the question – in this case, for example, the difference between 'why is going to university a good idea?' and 'why do *you* think that going to university is a good idea?'.

In other words, all of this illustrates that in any given situation there is a range of knowledge, or data, or propositions (why going to university might be beneficial) and a specific context to apply that knowledge to (why going to university might be beneficial for a specific individual). The former is, in fact, only useful in so far as it can be used to demonstrate the latter. It is not enough simply to *have* knowledge, you have to be able to *use* it.

What this also shows is how, in order to answer one question, we often have to consider the answer to lots of other, different questions.

TASK 1.2

To investigate this, let's look at question 1 again, and approach it from a different angle. This time, I would like you to try and think of as many questions as you can that might help to answer the question of why you think going to university is a good idea.

One example is the question that we looked at above: what is university for? What is its purpose for both the individual and the broader society?

See if you can think of five more:

1.
2.
3.
4.
5.

There are, again, a huge number of possible questions you could have chosen here. Some examples include:

- What should be taught, and should not be taught at university? Are some subjects more 'valuable', or 'useful' than others?
- Is university education a right, or a privilege? Should everyone have access to it?

- What should it cost to study at university? Should 'home' and 'international' students be charged differently?
- What values does the university stand for? Is it morally good, or morally neutral? Is it left-wing or right-wing?
- Who owns the knowledge that is produced at university? How should research be funded, and what should be researched?
- Is knowledge a commodity that can be bought and sold? Is a student a customer, or a scholar? Is there any difference, and can one be both at the same time?
- What will your degree/qualification be worth in 'real' terms?
- What skills actually make you useful in 'real life'? Will that be the same in 20 years' time?
- Has technology changed the status of knowledge?

Your answers to all of these questions will shape your overall answer to why you think university is a good idea. It is also possible to see something like a hierarchy of questions emerging, with our original question – why is going to university a good idea? – at the top, and the others arranged below, in levels of varying importance. That hierarchy will look different depending on your approach, but it could look something like Figure 1.2.

The questions at level 2 are essential in order to provide a clear answer to the primary question. The questions at level 3 (and there could be lower levels) help to inform the answers to the questions at levels 2 and 1, but they are not vital. That is, we could not answer them and still build a satisfactory overall argument, but answering them will provide greater

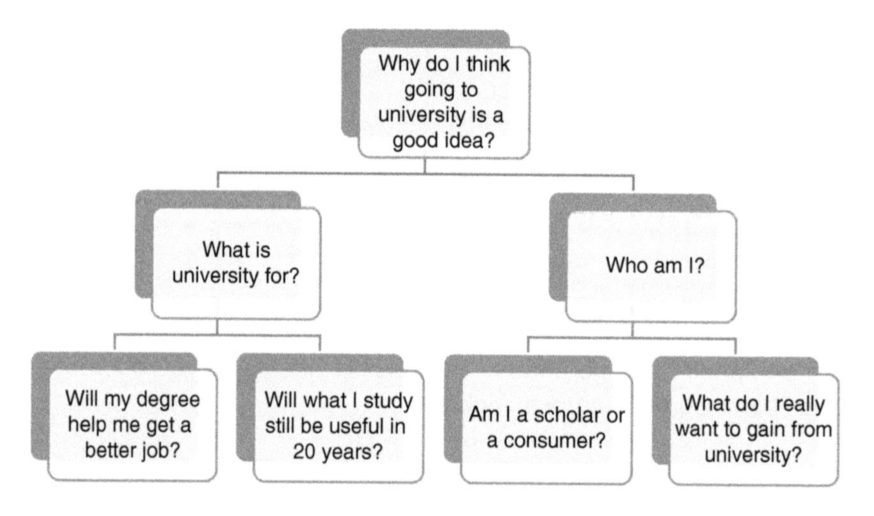

Figure 1.2 A hierarchy of ideas

depth and detail. When building any argument, it is important to keep track of which level you are engaging with at any given point, to make sure that your focus remains on the overall question, and not one of its subsidiaries.

The myth of the university

> A well-known scientist (some say it was Bertrand Russell) once gave a public lecture on astronomy. He described how the earth orbits around the sun and how the sun, in turn, orbits around the centre of a vast collection of stars called our galaxy. At the end of the lecture, a little old lady at the back of the room got up and said: 'What you have told us is rubbish. The world is really a flat plate supported on the back of a giant tortoise.' The scientist gave a superior smile before replying, 'What is the tortoise standing on?' 'You're very clever, young man, very clever,' said the old lady. 'But it's turtles all the way down!'
>
> Stephen Hawking, *A Brief History of Time*, 1988, p1.

The answers to all of the above questions provide us with a set of propositions that we can use to build an argument about why going to university is a good idea. What is also important, however, is that those answers are themselves based on arguments. To paraphrase the old woman who believed that the world stands on the back of a turtle – it is arguments all the way down!

In many cases it is straightforward to see what factors and decisions shape our answers to specific questions – we have already explored some of these. However, it is also vital to be aware that our responses to important questions are in fact often based on common collective arguments that we may not be aware of as arguments but rather think of as simply truths or self-evident ways of looking at the world.

One useful way to approach these fundamental views of the world that we all hold, is the idea of myths. Roland Barthes tells us that myths arise 'whenever things that [a]re culturally constructed [a]re given the appearance of being essentially entirely natural' (Docherty, 2015, p15), and the university occupies such an integral place in both individual societies and the whole contemporary global system that it is often quite difficult to see it as anything other than natural. That is, the importance of the university in both individual societies and the world system is such that it feels inevitable that it, or something like it, has to exist. That is, in fact, a myth.

Something like university has existed in the UK for around 1,000 years, and in other parts of the world for even longer, or for drastically shorter periods of time. The modern form of the university, however, is very recent – from the early to mid-nineteenth century – and the current globalised system incredibly so. It is important to remember this, so that what is contingent does not appear timeless.

There are many myths that surround the university, including the knowledge for knowledge's sake myth (otherwise known as the myth of the ivory tower), the getting rich myth (that university is a route to increased earning power and wealth), the route to somewhere better myth (that education is the means to transcend the limitations of your situation), the meritocracy myth (that university is the means to allow every person to realise their inherent potential), and so on.

Whichever myth you subscribe to (and as with most myths, you may not realise that there was anything that you subscribed to at all), the key here is that the university is *not* natural, no matter how much the myth-making makes it appear so. It is a construct, as all institutions are and all knowledge is, and seeing it as such enables you to take it apart, to analyse it and to remake and reshape it.

To explore this idea, there is one more question that we need to examine, and it is the question from level 2 in Figure 1.2 that we have not yet looked at – who are you?

Again, this is a question that is too broad to approach in its entirety, so it is necessary for us to consider some subsidiary questions to help us answer it. In this case, those questions are:

- What are your own assumptions about university? What does it represent in your life?
- Was it always expected that you would go to university?
- Are you the first in your family to go? Did your parents go to university?
- Did many of your peer group and friends go to university, or do they intend to?
- Was it a tough choice or an easy one?
    - Have you interrupted a career, or made a big life change to do this?
    - Are you staying close to (or at) home, or travelling far away?
- Does going to university guarantee you a job?
- Does it set you apart from other people? Or don't you really know?

Take a moment to think about these questions and take notes if you would like to.

Answering these questions, and others like them, will enable you to explore what assumptions you hold about university and how they affect your view of it. They will also help you to examine which propositions supporting your argument about why going to university is a good idea are, in fact, 'facts', and which are culturally constructed arguments that you have been socialised into believing without necessarily knowing why.

Why university is a good idea – revisited

All of the above shows us that even a seemingly straightforward question is not so straightforward once you examine it more closely. To answer one question, you often have to answer a number of other questions.

The question 'why are you at university?' is an example of a question that has lots of valid answers but no one right answer. This is because an argument only makes sense in a particular context, and that context shapes and limits what the answer to that question can be. As we will explore in later chapters, this does not mean that there are no *wrong* answers, simply that there is no definitive or final answer that can be reached.

The above also shows that arguments are individual in one sense, in that they reflect the conditions of the unique context in which they are constructed, but also that they are shaped by the collective conditions in which that argument occurs (e.g. the wider socio-cultural context). The university you go to will be part of a national system, and will also be shaped in some ways by an increasingly global system. You can make choices within that system, but you cannot go to a university outside of it, and your worldview, which will help to determine the choice that you make, is also shaped by the system within which you live.

What this shows us already is that to answer a question, and to build an argument in response to it, you have to understand:

- the terms of the question
- the assumptions behind the question
- the reason the question is asked – what is its purpose?
- the possible knowledge and evidence that is relevant to that question
- the different possible answers that can be based on that knowledge, and how they are shaped by different contexts and worldviews.

TASK 1.3

Let's go back to the beginning of the chapter, and reconsider question 1 again: why do you think going to university is a good idea? Go through the checklist above, and make sure you have considered each point properly. As you do so, think about all of the other questions that have been covered in this chapter and take notes.

Now, without looking back at the first sentence you wrote, once again write a one-sentence answer to the question: why do you think that going to university is a good idea?

A:

Does it match the first sentence that you wrote? If not, why not?

Why are arguments important at university?

Having considered our first question in detail, let us turn to the second – namely, why are you here, reading this book?

Again, we need to look at the terms of the question and establish what other questions we need to ask. We have already thought about 'you', but this time we also need to consider the idea of the academic skills book and, indeed, the author – that is, me. Why am I considered an authority worth listening to on the subject of arguments, or indeed universities? Should I be?

What is implied in the relationship between me, this book and you? Is it a straight transmission of knowledge from me, to the book, to you? Do I control the meaning of this book, or do you? Or is it both of us, or neither?

Is there a lack in you, a deficit, that I can help to fill, or solve with this book? Are you an empty vessel, to be filled? Or is the model closer to that of a discussion between equals, with whatever meaning you gain from this book produced by a conversation between us and the ideas that we discuss here?

These are questions that it pays to ask – even if it is perhaps uncomfortable, for both of us.

Another question that is nesting inside this is – why are arguments important at and to university? Important enough for this book to be published, and for you to want to read it?

There are a number of reasons, some of which are implicit to what we've already discussed. Some of the answers that I'm going to give here are also fundamentally based on a certain conception of the university – that is the 'global' university that is based on the 'Western', or at least Euro-American model (I include Australia in this). This is not solely English speaking, but English does predominate, and the majority of academic articles are published in English[1] – which again, is something that we might think about differently in light of Ngugi's argument about the colonial uses of the university discussed above.

As one way of answering this question, let us look at some examples of the most fundamental way in which student work is marked at university – that is marking criteria.

What gets you a 'good' mark at university?

One way to consider why arguments are important at university is to consider the question – how is student work judged in that context? In the current UK higher education system, a 'good' degree is defined as one where the student has achieved an overall mark of over 60, or a 2:1. Degree classifications work differently in different countries, but the lessons that we can learn from looking at how boundaries are set within an individual system are useful wherever you are studying.

TASK 1.4

Look at the following two examples of undergraduate mark schemes from UK university departments. In each case, see if you can identify any key differences in the criteria for a mark above 60 and for a mark below 60. Highlight any terms or phrases that seem particularly important.

Example 1 (from a department in the Humanities)

70+ Outstanding work which answers the question comprehensively, shows real understanding of the topic based on wide reading, clearly demonstrates powers of critical analysis and originality of thought. Displays skilful command of language, economy, precision and clarity in expressing difficult ideas.

60+ A comprehensive answer which is well-argued and logical. Demonstrates good understanding of the reading material and does not rely on lectures alone. Shows some independence of thought and critical analysis. Well constructed and fluently written.

50+ A fairly good answer, less well constructed and with less evidence of background reading. May be derivative or disorganised in its presentation or fail to address an aspect of the question.

Example 2 (from a department in the Social Sciences)

70–80% – A first-class essay shows a clear command of material, arguments and sources. It will show a clear understanding of underlying principles and a use of those principles in answering the question ... Where appropriate it will utilise empirical material to illustrate theoretical points. The essay will show independence of judgement.

60–69% – An upper-second-class essay shows a good knowledge of material, arguments and original and secondary sources. If it is in an empirically oriented subject it will show some relation between that material and appropriate theoretical and conceptual frameworks. If it is in a theoretical subject it will show some grasp of principles and development of argument ... The essay will make a clear point or points and show some critical acumen.

50–59% – A lower-second-class essay shows a basic, clear and generally correct knowledge of material, arguments and sources, particularly original sources. It will correctly summarise empirical or theoretical material, show some understanding of the material and its importance and draw reasonably appropriate conclusions.

There are many important points to consider here, not least, in Example 2, the repetition of the word 'argument'. In each case, however, the key terms to consider in the 60+ brackets are '*critical*', '*original*' and '*independence*', while in the 50–59 bracket, the key terms are '*basic*' and '*derivative*'.

'Critical' is another word, like 'argument', that means something different in an academic context. In everyday usage, 'critical' tends to have negative connotations, and to mean finding fault with something or someone. In a university setting, 'critical' is a positive term, and refers to active engagement with and analysis of thought and ideas, and to making evaluations of and judgements about them. These judgements can just as easily be positive as negative.

In other words, being critical at university is about questioning. Critical thought analyses and interrogates ideas rather than simply accepting them, and asks, for example, why information has been presented the way it has, why certain evidence has been chosen and certain evidence omitted, why a question has been phrased the way it has, and what assumptions reveal who is asking the question and what their purpose is, and so on.

Work that does not demonstrate these qualities falls into the categories of 'basic' or 'derivative' described above. This sounds very negative but is in fact better thought of as a qualitative difference rather than a value judgement. 'Derivative' simply means derived from, or based on something else, and is not inherently a bad thing. Work that falls into the 50–59 categories in the two mark schemes above would be work that showed a good understanding of everything that the student had been taught, but did not demonstrate anything outside of that, or the ability to apply or adapt that learning to different situations.

In this context, the word 'original' can seem intimidating, but it need not be. Here, the notion of originality is perhaps better equated with that last key term 'independence'. The above mark schemes are not expecting students to come up with ideas or answers that no one else has ever thought of – the criterion for a PhD is 'a new and significant contribution to knowledge', and that is certainly not expected at undergraduate level.

Rather, what is being looked for here is the contribution of the student, and their ability to use and engage with the knowledge they have been taught. This can ultimately be quite simple, and choosing to emphasise certain points rather than others, identifying gaps or weaknesses, or understanding which theory is the best way of explaining a particular issue can all be examples of critical, original or independent thinking.

Another way of thinking about the above is to say that work that simply describes or explains the ideas of others, no matter how well written or explained it is, can never get above a certain mark at university. This demonstration of having learnt and understood material is often enough to gain the very best marks at lower levels of study (for example, at school), whereas at university, the ability to understand, and learn, large amounts of complex ideas is a given – it is the expected norm.

This can be one of the most difficult transitions for students to make when beginning their university studies – the standard that was previously enough for excellence is now seen as the bare minimum. We will explore this further in Chapter 9.

TIP

Make sure that you familiarise yourself with the mark schemes that are used at your university, both for your overall course and for individual assignments. If you want to do well, it is vital that you understand the standard against which your work is going to be judged.

Arguments and criticality

A good understanding of arguments and how to build them is a vital part of achieving a critical approach, and is the way in which learning and acquiring new knowledge can be turned into the original or independent.

In other words, arguments are important because at university it is not just about learning knowledge but about learning how to *use* that knowledge. It is also about understanding how that knowledge is produced. One of the jobs of university is to produce new knowledge, to prove old facts wrong and create new ones, and while facts are not (necessarily) arguments, arguments arise when we need to say what facts *mean*.

As we have seen in this chapter, answering any question involves gathering and deploying a variety of points and using them to create a series of propositions that lead to a clear conclusion. To do this well requires the ability to identify arguments in the work of others, and to interrogate and evaluate those arguments. Argument is thus both a product and a process:[2] it is the answer, and the work of answering. Arguments produce answers, but also further questions, which in turn creates new arguments – and so on.

All knowledge is produced in this way. The example mark schemes given above related to the humanities and social sciences, but the ideas being discussed here are equally applicable to the sciences. For example, there is a wide range of facts that could be identified as to why a coral reef was declining – including tourism, fishing activity, increasing ocean temperatures, or acidification and pollution. Identifying which factor most urgently needed to be addressed and why would, however, require an argument to be constructed.

If arguments are all about questions, and criticality is about questioning, then we have, to an extent, come full circle, and can see how closely interlinked the two are. This constant iteration of argument and question, question and argument is the engine upon which the work of the university rests, both for students and for academics, and is the way in which knowledge is produced and reproduced.

In order to succeed at university, then, it is not enough to simply learn, it is vital that one learns to argue.

Summary

Let's summarise the key points so far:

- An argument is an answer to a question or position on an issue supported by propositions or reasons.
- Facts are not arguments, but arguments explain what facts mean, and how they can be applied to specific contexts or situations.
- Arguments are important at university, because students are judged not just on their ability to acquire new knowledge, but on their ability to interpret and apply that knowledge.
- Arguments are a key part of an active and critical approach, which requires and enables the ability to identify and interrogate arguments, as well as to build them.
- Arguments are shaped not just by facts and evidence, but by contexts and theoretical approaches.
- In order to succeed at university, it is important to have a clear understanding of what you want to get out of your studies, why university study is important to you, and the standards against which your work will be judged.

FURTHER READING

Burns, Tom, and Sinfield, Sandra, *Essential Study Skills: The Complete Guide to Success at University*, 4th edn (Sage, London, 2016).

Collini, Stefan, *What are Universities For?* (Penguin, London, 2012).

Gill, Jedd and Medd, Will, *Get Sorted* (Palgrave, London, 2015).

CHECKPOINT

Before moving on to the next chapter, let us revisit the question that I asked you at the beginning of this one – what do you want to get out of reading this book?

Try and write at least three clear goals that are achievable and realistic.

1.
2.
3.

NOTES

1. For an influential discussion of this topic, see Altbach (2007). For a discussion of how the promotion of English 'as the language of science, knowledge and internationalisation is an ideological act', see Li (2022, p9).
2. This phraseology is borrowed from Ramage et al. (2001, p9–10), although for a slightly different purpose.

2

Types of argument

There are many ways of classifying arguments into different types. One of the most fundamental differences in approaching this depends on whether you are looking to study arguments in themselves or are looking to understand arguments as they help to perform specific functions – for example, studying a subject at university or undertaking tasks as part of that study.

The purpose of this chapter is very much the latter, but in order to do that, it is helpful to first understand the different ways in which arguments themselves are discussed, even if only to provide us with a language that we can use to help us consider what university study requires.

Classical and formal logic

The study of argument has a long intellectual history, with its beginning most often located in the work of classical philosophers such as Aristotle. Aristotle's study of logic has been hugely influential in the modern Western philosophical tradition, and while what is commonly referred to as 'Aristotelian logic' often has only a tangential relationship with his actual writings, his thought remains foundational.

We use the word 'logic' or 'logical' in everyday speech to talk about something that is sensible or rational in its approach to understanding the world. However, *formal logic* is actually an attempt to scientifically approach how propositions and arguments – in other words, what we can say about the world – are formulated, tested and demonstrated. The study of formal logic (which has come a long way since Aristotle) is often a key introductory element of philosophy degrees, and it is an approach which, particularly in the USA, is often used to underpin writing courses at university.

This approach assumes a universality of logic which is, as we shall see in a moment, problematic, not least in terms of helping us to learn how to build good arguments in different disciplinary contexts. For our purposes here, there is, however, one pair of terms that is key to the study of logic, from Aristotle onwards, which will be useful to explore, and that is *deduction* and *induction*.

Deductive and inductive reasoning

Let's start with a definition of each of these terms:

- 'induction is the process of arriving at a generalised conclusion on the basis of observation' (Bonnett, 2008, p25)
- 'deductive reasoning: spelling out whatever conclusion follows logically from your premises without reference to any external information' (Chatfield, 2018).

Induction, then, is the process by which we make general statements about the world based on particular things that we have observed, while *deduction* is a process used to test arguments and see if the different parts add up to what we think they do. Inductive reasoning makes predictions about what will happen, or about what we don't yet know, based on what we do already know. Deductive reasoning is how we check that what we know is correct.

Another way to think about this is that inductive reasoning involves a move from the specific to the general, while deductive reasoning is the other way round, taking a general proposition and applying it to a specific situation.

Take a look at the following and see whether you can work out which ones are examples of inductive and which of deductive reasoning.

1. Every swan I've ever seen is white. Therefore all swans are white.
2. Fraud is against the law. The minister has committed fraud and so will be convicted of a crime.
3. You must have a visa to enter the country. I have a visa so the border guard will let me in.
4. There has never been a black Prime Minister of the United Kingdom. Therefore the next Prime Minister will be white.

(Answers: 1 and 4 are inductive arguments; 2 and 3 are deductive arguments.)

These very simple examples show us that this can be a useful way of thinking about arguments. However, we don't have to think about these examples for very long to see that there are also fairly obvious problems here.

TASK 2.1

Take a look at the four examples again and think about any potential issues you can see with each. Do you think they are good arguments? How would you challenge or disprove them?

1. ....
2. ....
3. ....
4. ....

In considering these arguments, there are a few more terms that we can introduce to our discussion.

1. Firstly, the argument that all swans are white is one that would be very convincing to many Europeans, as swans in Europe tend to be white. However, there are plenty of swans elsewhere in the world that are not, and because of this the 'black swan' has become a famous example of how inductive reasoning is open to *falsification*. That is, you only need to find one swan that is not white to prove that this argument is wrong, no matter how many white swans you have seen before. The philosopher *Karl Popper* is most commonly associated with this idea, and how it is used to demonstrate the importance of falsifiability to the rational and scientific process.

2 and 3. Examples 2 and 3 seem fairly straightforward at first glance, and yet even the most superficial examination leads us to realise that they are not. Not all ministers (political or religious) who commit fraud are convicted, and not everyone who fulfils all the conditions of entry to a country are allowed through by border guards. There are a few things that we need to consider here.

With deductive arguments, if the premises logically lead to the conclusion then the argument is considered to be *valid*. Both these arguments would appear to be valid, in this sense. But this judgement is made only by comparing the premises to the conclusion – it does not compare either the premises or the conclusion to the outside world, so we don't know whether they are true or whether there is missing information.

If the premises of the argument *are* also true, and the conclusion is therefore also true, then the argument is described as being **sound**. In these examples, we know that in many (if not most) legal systems in the world, powerful individuals such as politicians are able to escape the results of their actions, even if they were criminal. Equally, we know that border guards do not necessarily only consider visas when deciding who to admit to a country. Therefore arguments might be *valid*, but they are not necessarily *sound*.

4. The final example illustrates the fundamental *problem with induction*. That is, no matter how likely we think something is based on previous experience, we can't ever know for sure that it will be true next time. When I wrote this exercise, Boris Johnson was Prime Minister, and whilst the prediction that the next Prime Minister (Liz Truss) would be white was correct, the next incumbent was Rishi Sunak, which shows how difficult it is to make such forecasts.

All of these examples also show us, first of all, the importance of being careful about how we phrase our arguments and the language that we use. We could word these examples more carefully to make them more specific – the minister is *guilty of* a crime, for example, even if they are not *convicted*, and the border guard *should* admit the individual even if he *may* not. The next Prime Minister is *very likely* to be white, even if the UK has finally had a non-white head of government.

This shows how important it is to be *precise* and *accurate* in our language – in everyday situations it's fine to be approximate, but if you want to say something in an academic context, it's vital to say exactly what you mean.

Equally, in each of these cases, one obvious way to produce a better argument is to give more detail, or to add more premises, to put it in logical terms. That is, it is not that that these arguments are wrong, but rather that they are not *universally* correct – we need more information if we are to judge a specific case. If the individual is white, has a British passport and no criminal record, then argument 3 is much more likely to be correct. Equally, if the minister has committed their crime in a country with a strong rule of law, their conviction is much more likely.

This is of course true, but in a way it is not very helpful. Writing courses (or academic skills books) tend to use simplified examples like this while ignoring the fact that this is not that useful for building more complicated arguments in an academic context. There, while it remains necessary to make these simplifications at points, it is vital to remember that they *are* simplifications and not adequate representations of the complexity of the world. It is also important to remember the difference between discussions that take place in the abstract – i.e. in general terms, and not about specific examples – and discussions of concrete events or data.

In academic study, having a sense of inductive and deductive reasoning can provide a starting point for thinking about arguments, but in the actual activities that you undertake, the two will often be mixed up together and difficult to disentangle. Inductive reasoning, for example, can be used to build a hypothesis, which is then tested by deductive reasoning – whether using empirical data or another means of falsification. This is the basis of modern science and rational thought, and is the best means we have for achieving objectivity or something close to it. However, a key level of ambiguity needs to be recognised here and is often effaced. Logically, if the premises are true in a deductive argument, then the conclusion must also be true. But our premises are in themselves often based on inductive reasoning and further chains of premises – or they are based on coming up with something that we can test in order to prove something else (as with answering a simpler question or the hierarchies of questions in Chapter 1).

The key with arguments, whether inductive or deductive, is that we must resist the urge to make them definitive and eternal. This is true in every area of study, whether the humanities, social sciences or sciences. It is tempting to think of science as dealing solely in fact and objectivity, but 'science is a betting system, not a belief system' (Butterworth, 2011), and it in fact rests on the notion that premises must always continue to be tested. Stephen Hawking sums this up neatly:

> Any ... theory is always provisional, in the sense that it is only a hypothesis: you can never prove it. No matter how many times the results of experiments agree with some theory, you can never be sure that the next time the result will not contradict the theory. On the other hand, you can disprove a theory by finding even a single observation that disagrees with the predictions of the theory. (Hawking, 1988, p10)

This is a very good description of both the idea of falsification and the problem with induction, but it also tells us something even more fundamental. For Hawking, 'a theory is just a model of the universe, or a restricted part of it ... it exists only in our minds and does not have any other reality (whatever that might mean)' (1988, p9). Both inductive and deductive arguments, in other words, involve necessary simplifications, and restrictions of what is being considered – they are representations ('models') of the world and not the world itself.

As we have already seen, logical deduction considers only the elementsathand, and can be rendered invalid and unsound by the introduction of new elements, the consideration of excluded factors or comparison with reality. Induction is always open to falsification, and while it is possible to keep the two separate in theory, it is impossible in practice.

What this tells us is that one of the problems with focusing too closely on models of formal logic is that you end up focusing too closely on the system that you have created, or the examples that you have come up with, and not the real world that you are attempting to understand. We let the model stand in for the universe and forget that it is a model.

It also shows you not only that you could get very involved in arguing about arguments if you wanted to, but also that this type of study is very helpful if you want to think about how to study arguments, but in order to build, use and deconstruct arguments for your own purposes at university, it has its limits. There are some useful questions we can take from the idea of inductive and deductive reasoning though, to help us when we are either trying to understand arguments, or build our own. These are:

- Does the conclusion follow from the points used to support it? (Is the argument valid?)
- Are there other possible conclusions that could be based on these points? (Is this the only possible interpretation of these facts?)
- Is there any information or evidence that has not been considered? (What's missing?)
- Are all the points used to support the conclusion true, or correct? (Is the argument sound?)

TASK 2.2

Let's test this with an example. Look at the following extract from some marketing material published by the University of Essex in the UK in 2012 and consider the four questions above.

RESEARCH EXCELLENCE

Essex was rated ninth nationally in the last Research Assessment Exercise for our outstanding research, with most departments, schools and centres rated 'internationally excellent', meaning our students are taught, supported and supervised by world leaders in their fields.

Figure 2.1 Marketing material published by the University of Essex, 2012

The overall argument here is clear. The aim is to convince students that studying at the University of Essex is a good idea, and the claim is that if they do so, they will be 'taught, supported and supervised by world leaders in their fields'. This is supported by the fact that the university was ranked ninth in the UK for the quality of its research and that 'most departments' were rated internationally excellent'.

But does this argument stack up? First of all, let us check whether it is *valid* – that is, do the propositions logically lead to the conclusion? On the surface, it seems convincing enough, but the answer is no. Being ranked 'internationally excellent' is not the same as being a 'world leader', and the fact that there are eight universities in the UK ranked higher already calls this into question.

There is also some information here that is not given, or at the very least, that is sidestepped. Saying that *most* departments are ranked internationally excellent only means that more than half received that ranking – and given that this is taken from a piece of marketing, it is likely that if the figure was much higher than that, then the claim would have been made more specific (e.g. '90% of departments ...'). It is also not made clear which departments are ranked this way, and which others received lower rankings. Those producing research are also not necessarily those teaching or supporting students, which further complicates matters.

It could be, therefore, that this claim is true – and students at the University of Essex are taught by world leaders in their field. But we cannot know if this is true based on the evidence given here, and it could just as easily be the case, based on these same points, that the majority of students at the University of Essex do not receive world-leading teaching.

This is a useful example of how exploring whether an argument is valid and sound can be helpful, but more important, probably, is what it tells us about how suspicious we should be of university marketing departments.

Toulmin and Rogers

Before we move on to consider where arguments occur in a university context, there are a few more terms that we need to familiarise ourselves with. These relate to two other names that commonly recur when discussing the topic of arguments – Stephen Toulmin and Carl Rogers.

Toulmin provides us with a useful set of terms for talking about the building blocks of arguments while Rogers provides a means for thinking about how to resolve conflicting points of view and arrive at a consensus position. Let's consider both here.

The Toulmin model

British philosopher Stephen Toulmin laid out his model for argumentation in his 1958 book *The Uses of Argument*. He moved away from formal logic, and instead built an approach based on the idea that all claims and premises are contestable, and therefore need to be fully justified if they are to be successful (Toulmin, 2003, p90).

Toulmin proposed six necessary aspects for a good argument:

Claim: what you want to argue.

Data: what grounds your claim is based on – the evidence that supports it. For Toulmin, the question here is 'what have you got to go on?'.

Warrant: the justification that explains how the claim relates to the grounds. Toulmin's question here is 'how do you get there?' – in other words, how did you get from the data to the claim?

Backing: support for the warrant.

Rebuttal/reservation: any counterarguments to the claim or any restrictions that can legitimately be placed on it.

Qualifier: the degree of certainty attached to the claim. If you look back to the discussion of the four examples on the previous pages, you will see lots of examples of qualifiers expressing degrees of certainty, or how sure we can be about an argument. These include words and phrases like 'will', 'might', 'very likely', 'often', 'may' and so on.

It is easy to see how this builds on the discussions of formal logic above, and even corrects some of the issues that were identified. It is equally clear that you could usefully apply this model to the examples we discussed previously. For example:

- *Claim*: the border guard will let me in.
- *Data*: I have a visa.
- *Warrant/backing*: the law states that if I have a visa I will be allowed to enter the country.
- *Rebuttal*: there are many other possible reasons why a guard might refuse to admit someone, including criminal convictions, immigration restrictions, prejudice and so on. The visa is just one condition among many.
- *Qualifier*: the degree of certainty depends on the characteristics of the individual, and the immigration system of the country in question.

Most of these terms are fairly straightforward and are everyday words being used without any real special or technical meaning. Indeed, you can see how many of them have already been used in this book so far, in one form or another. This, at least in part, serves to demonstrate how well established Toulmin's model has become, even if you have never heard of it by name before.

The exception to this is probably the notion of the *warrant*. For Toulmin, this is what shows that 'the step' from the data 'to the original claim or conclusion is an appropriate and legitimate one' (Toulmin, 2003, p91).

Toulmin argued that the data we use to support our claim is usually explicit, while the warrant is usually implicit. That is, the evidence that we use to support our arguments is normally clearly stated, but the logic that we use to connect the two is usually assumed or thought to not need any explanation.

There are a number of different types of warrant. These include:

- *Generalisation*: what can be seen in a sample is also likely to be true of the whole.
- *Analogy*: that the relationship between the data and the claim is very similar to another example, with which it shares many similar properties, and so is also true here.
- *Causality*: that one thing causes another.
- *Authority*: a claim is correct because an authority confirms it. This could be the law, as in the example above; it could be academic or scientific authority; or it could be that a powerful or eminent figure holds this to be true.
- *Principle*: a rule or value that is generally agreed upon.

TASK 2.3

Take a look at the following examples and see if you can identify which type of warrant is being used in each.

1. This person's asthma is caused by smoking, because smoking causes health problems.
2. A child is not responsible for their actions, therefore they cannot be tried for a crime.
3. This shampoo cures dandruff, as 87 of 120 people in a trial said that this shampoo improved their dandruff.
4. I know that Chinua Achebe is a great author because my lecturer told me so.
5. Gambling addiction is like drug addiction, and can therefore be treated using similar methods.

(Answers: 1 = causal, 2 = principle, 3 = generalisation, 4 = authority, 5 = analogy.)

Toulmin recognised that it was not always easy (or indeed possible) to separate the warrant and backing from the grounds used to make a claim, but argued that the distinction remains useful. The notion of the warrant, or the *justification* that we use to support how our claims follow from our evidence, is a key one that we will return to both later in this chapter, and in Chapter 3.

For our purposes, it is again clear that the focus on how arguments themselves work has drawn us into theoretical complexity that is not necessarily helpful in everyday, practical terms. Exercises on the Toulmin model often only train us to build Toulmian arguments, or classify elements according to Toulmin's terminology. This is not necessarily immediately transferable into other activities – academic or otherwise.

In order to make this more useful for us, then, let's use Toulmin's model to set ourselves several key questions that we need to answer in order to build, or understand an argument:

- What is the claim being made here? What is being argued or what am I trying to argue?
- What evidence is given or do I have that this claim is true?
- What assumptions or logic are being used to make this argument? How is this being justified?

- What are the possible counterarguments to this claim? What basis could be used to argue against it? Are there any limits or restrictions to what is being argued that haven't been considered?
- How certain can we be about this argument?

TASK 2.4

Look back at the extract from Ngugi's (1986) *Decolonising the Mind* on p6 and see if you can identify all of the elements of Toulmin's model, using the questions above to help you. Some will be easier to work out than others.

Rogerian argument

The final model that is often discussed or taught in the study of arguments is the *Rogerian argument*, named after the psychotherapist Carl Rogers. This form of argument is all about finding common ground between arguments and counterarguments, and finding a compromise position, or synthesis, that can be accepted by both sides.

Rogers argued that the biggest barrier to communication, and thus to getting others to accept our arguments, was our emotional investment in our own position – and, by extension, the emotional investment of the *other* person in *their* position. In order to get anyone to truly change their mind, you need to do more than just disprove their points, or demonstrate that their conclusions are invalid or unsound. You need instead to 'listen with understanding', to demonstrate that you comprehend their position and the worldview that underpins it. Rogers (2017) calls this 'empathic understanding', which involves 'understanding with a person, not about him' (p131).

We will revisit this conception of communication and argument both in Chapter 3 on theoretical frameworks and in Chapter 8 on discussions in class. But for now, let us look at what Rogers calls a 'little laboratory experiment', which you can use to understand what he's talking about.

> The next time you get into an argument with your wife, or your friend, or with a small group of friends, just stop the discussion for a moment and for an experiment, institute this rule: 'Each person can speak up for himself only after he has first restated the ideas and feelings of the previous speaker accurately, and to that speaker's satisfaction.' You see what this would mean. It would simply mean that before presenting your own point of view, it would be necessary for you to really achieve the other

> speaker's frame of reference – to understand his thoughts and feelings so well that you could summarize them for him. Sounds simple doesn't it? But if you try it you will discover it is one of the most difficult things you have ever tried to do. However, once you have been able to see the other's point of view, your own comments will have to be drastically revised. You will also find the emotion going out of the discussion, the differences being reduced, and those differences which remain being of a rational and understandable sort. (Rogers, 2017, p131)

Notice how the rule involves not just being able to restate the other person's ideas, but also their feelings, and being able to do this *to the other person's satisfaction*.

This is very different from what many of us tend to do when we listen to an argument that we don't agree with, which is to ignore the other person's feelings and to look for the weaknesses in their ideas. Attempting to inhabit someone's way of seeing in this way is challenging, but potentially very powerful.

TASK 2.5

Think about the last time you disagreed strongly with someone over an issue. This could be about something academic or something everyday, like which football team is better or the best way to make jollof rice. Try and institute Rogers' rule and see if you can think back and understand the other person's point of view well enough that you'd be able to explain it back to them in a way that they would be happy with.

Would this change your response to their argument? Would you be more likely to change their mind having done this? Might it change *your* point of view?

This approach is good for resolving conflicting points of view, but it is also a very useful approach to take when critically evaluating an argument in an academic context. We will return to that in Chapter 3, but first, let us look at some of those academic contexts and the arguments that appear in them.

Arguments at university

One of the problems with focusing too closely on formal logic when helping students learn how to build arguments is what John McPeck (1981) calls the 'philosopher's fallacy' (p8), or the idea that thinking critically is only about logic that can be learnt as a set of universal rules, which are then applied to any situation. This approach underpins the way writing, academic skills, and how to structure arguments is taught at many universities, particularly in the US and UK. But the idea that a generic set of skills can be taught that will then smoothly transfer into students' disciplinary contexts looks increasingly like it just isn't true (see, for example, Bruce, 2020).

This is not to dismiss the importance of models like Toulmin's (or indeed Aristotle's). As we have seen so far in this chapter, such models can be useful as long as their limitations are always kept in mind. Rather, it is to suggest that an argument is always born out of and relates to a specific context,[1] and that in order to make a good argument it is not enough to know what a good argument is in the abstract – you have to know what a good argument is *in a particular subject*: that is, according to what Swales and Feak call 'the standards of judgement of that field' (Swales and Feak, 2012, p328).

Consider these extracts from advice on arguments given by two university departments:

> *History*: Effective argument depends on evidence to support its points and on logical exposition. If you say something with which a reasonable person might disagree, clinch the point by citing examples ... Remember that most essays are, in large part, concerned with explanation: demonstrating why something in the past happened.
>
> *Philosophy*: Equally important is the skill of argument. This is a matter of stating clearly a view or position, and then of stating reasons why someone should adopt this view or position if he or she does not adopt it already. The standards for good argument are much more explicitly studied in philosophy than in other subjects.
>
> (University of Essex, 2016)

What is the difference between the two? Do you think that the two departments would be looking for the same thing in a 'good argument'?

We will return to the question of how disciplinary differences shape the way arguments are built in more detail later but for now it is worth noting a couple of points. Firstly, the emphasis in the History department's guidance is on *evidence* and *explanation*, while the emphasis from the Philosophy guidance is on the *logical construction* of the argument itself. It is also worth noting how the Philosophy department explicitly states that it has higher standards for judging these things than other subjects, which suggests both an adherence to the fallacy that there is one standard against which all arguments can be judged, as well as a certain sense of disciplinary self-importance.

If, however, it is not enough to learn a set of rules that claim to be universal, then how can you learn to build good arguments when studying particular subjects? It certainly makes things more difficult, not least for a book like this. What, then, can we do?

In order to do well at university, it is not enough to understood different models of argument – and arguably it is not even necessary at all. Instead, it is more useful to have a clear understanding of what it is you will be asked to produce – that is, what assignments you will be judged on, whether written or otherwise – and what the purpose of those assignments is. The argument is then the means by which you will achieve that purpose, and will be bound as much by the rules of the type of assignment as of any model of formal logic.

Genre: a form or type of writing with an agreed or standard set of rules

The approach we will take here, therefore, is twofold. Firstly, we will look at the different **genres** of assignment that students are required to write when studying different subjects, and secondly what trigger questions can be used to work out the standards that the student is expected to meet, and the rules they are expected to follow when writing those genres.

Types of writing

There are numerous genres that students will encounter at university where they are expected to build arguments. These will vary greatly depending on the subject being studied, and include, but are not limited to:

- *Essays:* extended pieces of continuous prose in response to a question. These are the most common form of assessment in the humanities, but are used across all disciplines. Be careful, however: exactly what is meant by an essay varies from subject to subject and country to country.
- *Research/lab reports:* Research reports have a standard structure and are used to 'write up' lab experiments, or other types of research. Usually, a research question or hypothesis will be established based on a review of existing literature (see below). A method of answering that question or testing that hypothesis will then be designed, producing results and a discussion of what those results mean. Among other functions, a report like this should allow someone else to repeat the experiment or research in order to verify it.
- *Literature reviews*: A critical survey of what else has been written in a particular field or area. In some forms of writing (e.g. a research report) this is

demarcated as an explicit section of the text, while in others it is an implicit and integral part of the whole (e.g. a humanities essay).

- *Case studies:* A particular event, situation, or example used to illustrate a category, theory or common situation. Although not always explicitly involving argument, the way in which case studies are framed and constructed (e.g. what information is included and what excluded) often involves a *de facto* argument, as we shall see in Chapter 3.
- *Short answer/problem questions:* Often used to test students' ability to apply a rule, principle or theory to a particular situation. In Law, for example, a student could be asked to respond to a case study by identifying which rule applies and how it should be applied (this is sometimes known as an IRAC question – Issue, Rule, Analysis, Conclusion). In Business, a case study could be given to which a student has to apply a theory of management or marketing, for example.
- *Reflective writing:* When students are asked to write about their own experience, either of a learning situation or of 'real life', how it has affected them and what action they will take based on that experience. Many students find this type of writing particularly challenging, as it involves switching between registers (e.g. objective and subjective, impersonal and personal), but it is an important element of many types of study.
- *Presentations/verbal response:* A presentation, or verbal response test, can take the form of any, or all of the above types, but there are key differences to the form these take when responding orally instead of in writing. In some educational cultures, having to respond to questions verbally, or 'defend' a position will be the primary form of assessment, and in some subjects, presentations and spoken tests are common. We will return to this in Chapter 8, but it is worth noting now that writing is not the only form that students will encounter arguments in.

TASK 2.6

Take a look at the list above and think about which ones you are expecting to encounter and be asked to produce in your studies. Which are you most familiar with and which do you expect to find challenging?

....

....

....

....

This list is a useful place to start, but it clearly only helps us to a certain extent. While each separate genre comes with its own rules of structure and expression, there are multiple types of argument that can be used in each, and those types of argument will not always be consistent across or even within disciplines. Some of what is happening in each genre will not even be an argument at all, and many elements will take the same form but be very different depending on the context.

In order to tackle this problem, consider the following two questions. Take notes on each:

- *What are academic arguments for*? Why are students asked to produce the various genres listed above, and what is the purpose of the various forms?
- *What does your discipline mean by 'truth'*? This is a deliberately abstract question, which I will not gloss any further here. It's a very difficult question to answer, but a very important one – see what you can come up with.

What are academic arguments for?

> Academic arguments are not about the subject or separate from it – arguments *are* the subject.

There are multiple answers to this question, but for our purposes here, let's think about two. The first relates to the overarching purpose of academic study – namely the production and transmission of knowledge. That is what academics in universities are for, to produce new knowledge, new ways of thinking, and to communicate those ideas, whether to other academics, to students or to the wider world.

As a student, your purpose is the same. This is an intimidating aspect of study at either undergraduate or postgraduate level, but it is an important one to be aware of. As a university student, you are a trainee member of the academic community, and the arguments that you produce are part of the same discourse as the arguments made by your lecturers. A 'subject' or 'discipline' only exists in what is written about it, in the conference papers that are delivered on it, in the seminar presentations and discussions that happen in university classrooms, and so on. Academic arguments are not about the subject, or separate from it – arguments *are* the subject.

Of course, as a student the 'trainee' aspect of your membership of the academic community is also important, and you are therefore also at university to learn about the subject you study. As such, the purpose of the arguments that you make is to *demonstrate to your lecturers what you have learnt* from their modules, in terms of both *knowledge and skills*. That is, you are asked to produce these different types of argument in order to show what you know, and also your ability to do things with that knowledge – whether that be the ability to apply existing knowledge to a new situation, to critique and analyse other research, or to write a coherent, valid argument in response to a specific question.

What does your discipline mean by truth?

At first glance, this may seem like a confusing question, as surely there is only one standard of truth? Particularly if you come from a scientific perspective, it is tempting to think that any attempt to undercut the notion of a universal standard of truth is a form of relativism that is fundamentally dangerous and responsible for many of the ills of the contemporary world, from fake news to 'post-truth' politics.

However, without needing to get into such fraught areas, there are certain questions we can ask to demonstrate how an interrogation of the notion of truth might be helpful. For example, if we rephrase this question as 'what does your discipline mean by *a good argument?*' it seems less immediately threatening, perhaps, but is ultimately a very similar question. Consider the extracts from departmental guidance above – what makes a good argument is different for a philosophy lecturer than for a history lecturer, and a big part of that difference is about what is the most important consideration in judging whether something is true. For one, it is about whether there is sufficient evidence for a position, for another it is about the logic that has been used to support a particular claim.

The notion of a 'good argument' is a construct that differs from context to context and culture to culture. Thus different cultures have different criteria for what makes a good argument and so do different educational cultures. Factors such as level of complexity (e.g. school vs university), social context (e.g. formal or informal), and subject of study all have an impact on how arguments are evaluated and to answer that question of what makes a good argument, we also have to consider the evidence that is being used. If you are a literature student, the evidence you draw on, for example, will be very different from the evidence that a life sciences student will be using.

The notion of the warrant can be useful here, as what is given as a 'good argument' or what is said by a discipline to be 'true' is often not just about data or facts, but rather a claim about what such data or facts can be said to mean, and the logic that is used to get from one to the other. Different subjects of study ask questions in different ways – consider, for example, how the question 'what is beauty?' might be approached by an art historian, a psychologist and an evolutionary biologist.

The study of how knowledge is produced and constructed is called **epistemology**. If, as above, we accept that the purpose of academic argument is the production of knowledge, then the epistemological differences (e.g. what evidence they use and how claims are verified) between different subjects are vital. These differences are ones that you will have to learn for yourself, but in order to learn them you have to be aware that they exist in the first place and know what questions to ask.

> **Epistemology:** the study of how knowledge is produced and constructed. Different subjects at university have different epistemologies – that is, they construct their answers to questions using different standards and types of evidence, and different structures of justification and logic.

What this means in practice is that as you progress your studies, think not just about what you are learning, but how what you are learning has been constructed. This applies on multiple levels: to the texts and sources that you are asked to read or engage with; to the assignments and tasks you are asked to carry out; and to the modules and courses themselves. In each case, you can ask yourself the following sets of *trigger questions*, to help make sure you are thinking critically about what you are doing:

Your own modules, courses and lectures:

- What is this module trying to teach me?
    - Think about both the knowledge and the skills that you are being taught. What is an end in itself, and what is being done to help you learn a particular way of doing things? For example, if you are being shown how to do a literature review, that is both a particular skill in

itself and an important aspect of training you to undertake longer pieces of research.

- Why has my lecturer designed the module in this way?
  - Is there a reason why the topics come in a particular order? What links can I make between the different sections?
  - How does this module fit into my course as a whole? How does it fit with other modules and what links can I make with other modules?

Evaluating sources and texts:

- What evidence is being used, and what is the logic used to interpret this evidence?
- What has *not* been included? What's been missed out and why?
- What sorts of questions are being asked?
- Why are these questions being asked, and in this way, and why are others not?
- How would another subject approach this question, and what might that difference tell me?

Building your own arguments:

- What is the context for my argument? Why have I been set this task or question, and how does it relate to what I have studied?
  - Another way of thinking about this is – why has my lecturer set this particular task? Why this task and not another one? What are they trying to show me and get me to demonstrate?
- What does a 'good argument' look like here?
- What am I going to include and what am I going to leave out? Why?

When both evaluating the arguments of others and our own, one key question to ask is '*so what?*'. That is – why does this matter? What is the purpose of this argument and what is it achieving?

TASK 2.7

Take a look at the following chart and see how much of it you can complete. Some parts have been filled in already to help you. Note that there is more than one possible answer in most cases, and some answers will vary depending on what subject you are studying.

Table 2.1

Genre	Evidence used	Purpose of argument – the 'so what?'
Essay	Depends on subject – e.g. historical evidence; source text quotations; government policy, etc.	
Research report		
Literature review		
Case study		To provide specific example of general category
Short answer/ problem question	Case study	
Reflective writing		

Answers

Table 2.2

Genre	Evidence used	Purpose of argument – the 'so what?'
Essay	Depends on subject, e.g. historical evidence, source text quotations, government policy, etc.	To answer a specified question; to explore a new way of thinking about an existing subject; to explain a complex idea or theory; to compare different perspectives on a question, etc.
Research report	Account of research/ experiment undertaken	To answer research question or prove hypothesis; to enable others to duplicate research
Literature review	Existing literature written about topic of area	To summarise relevant research in a particular area; to identify gaps in the literature, i.e. a question that hasn't been asked yet, or any area that needs to be looked into in more detail; to contextualise current piece of research and justify research questions

(*Continued*)

Table 2.2 (*Continued*)

Genre	Evidence used	Purpose of argument – the 'so what?'
Case study	Outline of event, occurrence, specific case	To provide specific example of general category; to show what a particular theory or category looks like in practice/'real life'
Short answer/ problem question	Case study	To apply a general theory or method to a specific case; to demonstrate the ability to put a particular method or approach into action; to demonstrate ability to apply theory to reality and perform key task (e.g. diagnose illness, solve engineering problem, advise on legal position, etc.)
Reflective writing	Personal experience	To identify potential improvements to personal, collective or general practice; to demonstrate understanding of theory by applying it to individual experience

The answers given above are indicative and not exhaustive, and some vary depending on the discipline in question. However, thinking in this way about what you are asked to produce as a student – and the sources you encounter – is a useful exercise in ensuring that you don't just think about what it is you are doing but how, and most importantly *why*.

In order to answer those questions fully, we need to have an understanding of the theoretical frameworks that underpin different disciplines, and ways of understanding the world, which we will look at in the next chapter. However, first, let us try and bring everything we have discussed so far together, and think about what we have learnt about the types of argument that appear in the different genres that a student will encounter at university.

Summary

- Academic arguments are the means by which knowledge is produced and communicated, and the way in which students demonstrate the knowledge and skills they have acquired in their studies.

- Formal logic can provide us with useful ways of thinking and talking about arguments, but learning how to build successful arguments at university is about understanding the specific standards and conventions of individual subjects and genres, not universal rules of logic.
- In order to understand what makes a good argument in an individual subject, we need to think about the criteria used to judge an argument in that subject, including the evidence that is used, the claims that are made based on that evidence, and what the purpose of those arguments is.
- The purpose of an argument is defined by the context in which it is made – that is, not just what the argument has to say about the topic or area, but the function that it serves within the particular genre or situation where it appears. In other words, why is this important – or so what?
- Academic arguments are not about the subject, or separate from it – arguments *are* the subject.

FURTHER READING

Chapter 1: Our Picture of the Universe from Hawking, Stephen, *A Brief History of Time* (Bantam, London, 1988).

Chapter 2: The Logical Structure of Arguments from Ramage, John D., Bean, John C. and Johnson, June, *Writing Arguments: A Rhetoric with Readings*, 5th edn (Allyn & Bacon, USA, 2001).

Rogers, Carl, 'Communication: its blocking and its facilitation', from *ETC: A Review of General Semantics*, Vol. 74, No. 1/2 (2017), pp129–35.

Toulmin, Stephen, *The Uses of Argument* (Cambridge University Press, Cambridge, 2003).

NOTE

1. This relates to a point that McPeck (1981) makes about critical thinking. I am also drawing here on Bruce's (2020) discussion of critical thinking in EAP.

3

Theoretical standpoints

Objectivity is a key aim of academic research, whatever the subject. Whatever knowledge academic study produces has to be impartial, impersonal, unbiased and rational, otherwise its claim to truth can be undermined by the fact that it is based on subjective opinion, personal belief or political and personal preference. Even where there is no claim made to impartiality, and the personal is a recognised part of the process, as with, for example, autoethnography or action research, the aim is still to contribute to and improve the overall body of objective knowledge, and these methods are seen as an exception to the rule rather than a challenge to it.

As we have explored in the first two chapters, academic knowledge is based on empirical fact and sound argument, and different disciplines have different methods for ensuring that facts are tested in the most objective way, and that the arguments constructed to explain what those arguments mean are valid and sound. In the sciences and social sciences, for example, there is the notion of the true experiment, which aims to ensure that bias or simple chance do not influence the collection and testing of data, while all disciplines seek to remove value judgements, support all points with evidence and balance points of view fairly rather than give undue weight to one position over another.

However, true objectivity is impossible, either in terms of how 'facts' in themselves are established or in terms of how arguments are built to explain those facts. Indeed, why would we need arguments at all if it was possible to be entirely objective or certain? The idea of objectivity suggests that it is possible to find a neutral position from which all other positions can be judged, and also that any absolute truth can be discovered/achieved. This is an idea that much of post-Enlightenment rational and scientific thought is based on, implicitly or explicitly,[1] even when its impossibility is also recognised.

However much we recognise the impossibility of completely objective or certain knowledge, however, it is always very tempting to believe that this is a problem that can be overcome, if only we try hard enough, or that we are much better at avoiding the potential pitfalls than others. The historian Richard Westfall puts this very well when discussing the experience of writing his famous biography of Isaac Newton:

> It is impossible to study history seriously in the second half of the twentieth century without acknowledging its subjective aspects. Like every other historian, I dutifully learned that lesson early and have never dreamed of denying something that seems obvious. My acquiescence was always conditioned by the silent proviso that subjectivity applied more to others than to me ... (Westfall, 1985, p188)

This might sound like an issue that is only relevant to subjects in the humanities, but despite our tendency to think of the sciences as completely objective, the same difficulties apply, and are in fact inherent to the scientific process. As Jon Butterworth (2011), a theoretical physicist put it, when discussing an experiment to discover if a theoretical particle (the Higgs boson) existed:

> ... scientific knowledge is about probabilities; it is provisional ... We should all know that science is a betting system, not a belief system. Near-certainty arises from a morass of uncertainty, it does not drop from heaven gift-wrapped. You never know, 100%. But you would be a fool to bet against a well-established scientific fact, be it gravity or the existence of quarks ...
>
> [After the experiment] the odds will shift either in favour of or against the existence of the Higgs boson. The Higgs (for or against) might make it to the 5% or 10% level: 5% chances turn up quite often in the real world. Eventually, the odds for or against its existence may be many millions to one. At that point we'll pretty much stop discussing that, and move on. Apples fall down, the sun comes back in the morning, the proton is made of quarks. These are all pretty certain at the billions to one level. (Butterworth, 2011)

To a certain extent, then, this impossibility of ever being truly certain about a fact is a moot point – as Butterworth says, there comes a point when you can safely ignore it. However, while we might be able to efface this in terms of the establishment of some facts (apples fall down, the sun comes back up ...), we cannot (and should not) efface it when it comes to saying what these facts mean, or the reason *why* those things happen – which is where arguments come in.

In order to explore what this all means for you as a student at university, this chapter will look at some of the reasons why it is impossible to be sure of what we know or achieve objectivity, before turning again to how knowledge is constructed and the theoretical frameworks which make that construction possible, so that we can make sure we build the best arguments we possibly can.

Cognitive biases and the problem with statistics

In large part, the problem with trying to work out what the final, objective answer is to any problem is that it is *us* trying to do the working out. In other words, it is a function of being human.

Human beings are meaning-making creatures, who have what the analytical psychologist Jung called a 'religious instinct' to find patterns and meanings in the world. While this instinct is behind a huge amount of human progress and achievement, it can also lead us to be overconfident in our ideas, or to see certainty where there is none. And this overconfidence is not (only) a result of individual failings, it is a function of how our minds work.

In his book *Thinking Fast and Slow* (2012), Daniel Kahneman discusses two main systems that underpin how we think. System 1 is automatic and very good at 'continually construct[ing] a coherent interpretation of what is going on in our world at any instant' (p13). We don't have to make any effort to do this, it just happens. System 2, on the other hand, is what we use when we have to think about complex, complicated problems, and we find thinking in this way tiring and difficult.

The problem here is that while System 1, which is what we use most of the time, in general does a good job at interpreting what is going on around us, it is also flawed and prone to mistakes. This is the key insight of Kahneman's (2012) research – that 'our minds are susceptible to systematic errors' (p10), and that these errors can be traced 'to the design of the machinery of cognition rather than to the corruption of thought by emotion' (p8) or anything else. The system works well most of the time, and that means that we don't notice when it does not.

One of the key difficulties we have is with thinking statistically. That is, we find it very easy to think in terms of cause and effect and in terms of metaphor, 'but statistics requires thinking about many things at once, which is something that System 1 is not designed to do' (p13).

In order to illustrate this, let's consider some examples.

TASK 3.1

Take a look at the following claims that are based on statistics and see if you can spot any problems:

1. 'You can make an estimate – one that is better than chance would produce – of how many children have been born into a Dutch or Danish family by counting the storks' nests on the roof of their house. In statistical terminology it would be said that a positive correlation has been found to exist between these two things' (Huff, 1991, p84).
   - o There is an old myth, popularised in Europe by Hans Christian Andersen in the nineteenth century, that storks deliver babies. Does this positive correlation mean that the myth is true?
2. 'The death rate in the US Navy during the Spanish-American War was nine per thousand. For civilians in New York City during the same period it was sixteen per thousand. Navy recruiters later used these figures to show that it was safer to be in the US navy than out of it' (Huff, 1991, pp80–1).
   - o If these figures were correct, were the recruiters right to argue that joining the navy was actually likely to make your life less dangerous?
3. 'Somebody once went to a good deal of trouble to find out if cigarette smokers make lower college grades than non-smokers. It turned out that they did. This pleased a good many people, and they have been making much of it ever since. The road to good grades, it would appear, lies in giving up smoking, and to carry the conclusion one reasonable step further, smoking makes dull minds.

   This study was, I believe, properly done: sample big enough and honestly and carefully chosen, correlation having a high significance and so on' (Huff, 1991, p85).
   - o Based on this study, is it correct to argue that smoking will make you get worse grades?

The first claim here is a classic example of the maxim that *correlation is not causation* – or in other words, just because two things are related does not mean that one causes the other. There are many famous examples of this, although my favourite is that there is a strong correlation between the number of films made starring the actor Nicholas Cage in the years 1999–2009 and the number of people who died in swimming pool accidents in those years.

There are a number of things going on here. Firstly, this illustrates the way in which, as Kahneman (2012) puts it 'people are prone to apply causal thinking inappropriately, to situations that require statistical reasoning' (p77), and it also illustrates how tempting it is to accept patterns like this and not do the harder thinking required. When it comes to storks bringing babies or Nicholas Cage films leading to drownings, it is not that difficult to realise that something might be wrong, but when the correlation is more tempting, as with the example of smoking in claim 3 above, it is much more difficult to realise that we need to interrogate the conclusion that System 1 is jumping to, and do some more detailed thinking.

Claim 2 is an example of *a sampling error* or of *false equivalence*. Again, it feels fairly obvious that there is a problem here – after all, it *can't* make sense that it is safer to be at war than it is to live in a city in one of the richest countries on Earth, can it?

To understand the problem here, we have to think about the two *populations* that are being compared. In the navy, we are likely to find young men, who are all fit and healthy, and who are provided with good diets and health care. In New York City, on the other hand, the population is both much larger, and also includes everyone from the very young to the very old, the sick and the infirm, and the very poor, who are unlikely to either live healthily or have access to healthcare. Comparing the two populations directly is therefore not possible, as they are not equivalent.

In example 3, the problem is again partly one of correlation not implying causation – just because smokers get lower grades, doesn't mean they are getting lower grades *because* they smoke. However, there are other issues here. The main one is that the many other factors which this group have in common are not considered. Taking up smoking is often related to stress, for example, and it could therefore in fact be that these students smoked *because* they got poor grades, not the other way round – even as more nests on roofs in claim 1 is most likely to do with the size of the house (which is because of the number of children that live there), rather than because those birds brought the children.

Another big issue here is that the conclusion fits an existing pattern. As is implied in the 'this pleased a good many people', there is a sense in which this is a desired outcome – people *want* it to be true – as it fits existing preconceptions. We know that smoking is bad for your health, so it seems likely that it could also be bad for your mental acuity. This tendency to be persuaded by new information or evidence that supports what we already think or know, and not to consider other explanations or dismiss conflicting evidence, is called *confirmation bias*.

In each case here, part of the problem is poor reasoning, or faulty logic, which should be fairly easy for us to be on guard for and therefore correct. However, the link to systematic errors in our thinking show how important it is not to be complacent about how pervasive such difficulties can be, not least because catching such errors requires us to consciously interrogate our ideas and step outside of our usual, automatic processes.

There are a large number of these errors discussed by Kahneman (2012), but for our purposes here, let's look at just two more.

The first of these is *substitution*, or the way in which 'when faced with a difficult question, we often answer an easier one instead, often without noticing the substitution' (p12). One example that Kahneman (2012) gives is that when people are asked to evaluate how happy they are with their lives, the question they will in fact answer is about their mood in that particular moment (p98). To further demonstrate this, psychologists have shown that if you ask people 'how many dates have you had recently?' before you ask them about how happy they are with their lives, their answers will actually be about the number of dates and will not consider any other factors (p101).

Kahneman (2012) refers to this phenomenon as *'WYSIATI'* or *'What You See Is All There Is'*. In other words, when we are interrogating something, we tend to only deal with ideas that are already to hand, and 'information that is not retrieved (even unconsciously) from memory might as well not exist' (p85). This *availability heuristic*, or our tendency to work with only what information is immediately available, combined with *confirmation bias* creates a tendency to be overconfident. As for our automatic, associative mind, 'it is the consistency of the information that matters ... not its completeness. Indeed, you will often find that knowing little makes it easier to fit everything you know into a coherent pattern' (p87).

In other words, our instinctual sense of whether an argument is correct or not will often depend more on how consistent it seems, both internally and with our existing views, and not how complete or sound it is.

Again, these systemic errors may seem relatively easy to overcome, especially if we pause and make the effort to concentrate and think in depth – to use our System 2, as Kahneman would put it. That is partly true, but it is very important to be aware of the ways in which the very mechanics of our thoughts can lead us to make errors of judgement and reasoning.

In a non-academic context, all of these biases and errors are part of what has led to current global issues with 'fake news' and conspiracy theories, whether these be about national elections or vaccines. From an academic perspective, we have already explored in Chapters 1 and 2 how academic research involves a necessary element of substitution, where an easier question is formulated in order to deal with a more difficult one – think, for example, of how we might actually be answering the question 'will going to university get me a better job?' when we think about the question 'what is university for?' (p18).

The key thing that this tells us is to be aware of our tendency to be more certain than we should be about the accuracy – and objectivity – of our understanding of the world. And this is not simply because of a failure to be sufficiently rational or being overly emotional – it is built into the very system of our thinking.

Assumptions

Even if we were able to fully insulate ourselves from any issues and make sure that we avoid flaws in our logic or errors in our thinking, we have an even more fundamental issue to deal with.

Consider the following:

Acting on an anonymous phone call, the police raid a house to arrest a suspected murderer. They don't know what he looks like, but they know his name is John and that he is inside the house.

The police burst in on a carpenter, a lorry driver, a mechanic and a fireman all playing poker. Without hesitation or communication of any kind, they immediately arrest the fireman.

How do they know that they've got their man?

You may find this old riddle straightforward – but if you didn't, you're not alone. Consider the following, which is a variation on the theme:

A father and son are involved in a horrific car crash, which kills the father and leaves the son in a critical condition. When he is taken to hospital for an urgent operation, the surgeon in the operating theatre says – 'I can't do it – this boy is my son!'

How is this possible?

In the first instance, the answer is that the fireman is the only man in the room – the rest are women. In the second, the answer is that the surgeon is the boy's mother.

In each case, it should be easy to work out the answer, and yet for many people it isn't. This is because, in **patriarchal** societies (and all current societies are patriarchal – the only question is to, 'what extent?'), we are conditioned to think of occupations in gendered terms, and so we automatically assume that surgeons, mechanics, lorry drivers and carpenters are male.

This might seem like a superficial point, but it is not. This is just one example of how assumptions shape the way that we see the world we exist in, and how we respond to questions about that world. If we are to build good arguments and understand the arguments of others, we need to be aware of the assumptions of others and, vitally, our own.

> **Patriarchy** is the term for a social system in which men hold the power and women are excluded from it or marginalised in other ways. All modern societies are patriarchal, to a greater or lesser extent.

Patriarchy is just one example of a theoretical framework that underpins our understanding of the world. Even if you do not consciously believe that men are superior, or that society should be organised to their advantage, these notions are historically and systemically embedded in the world system in a way that it is very difficult, if not impossible, to escape.

In order to explore this idea, let's look at some of the different types of framework and think about how they are relevant to studying at university.

Theoretical frameworks

There are many different ways of thinking about the frameworks within which knowledge is constructed, and many different ways of categorising them. For our purposes here, we will consider four – social and cultural, political and economic, disciplinary, and theoretical. These are impossible to completely disentangle – patriarchy could be discussed under each heading, for example – but it is useful to consider them separately.

Social and cultural

Society is a fairly straightforward term for us to understand – it refers to the aggregate of people who live in a given large grouping or community. Culture, although it is a word that is often used as if it is straightforward, is actually much more complex, and is indeed, as Raymond Williams

(1976) tells us, 'one of the two or three most complicated words in the English language' (p87). He never does tell us what the other two are, but he does provide us with two broad areas that are covered by this term. The first is to do with the 'way of life' of a group of people or in other words the shared habits, practices and values of a society. The second area covers 'cultural production', or in other words, all of the things that are produced by a group of people – whether that be objects, art, TV, music or food.

For our purposes here, we can largely stick to the 'way of life' side of things. Again, this might seem simple at first glance – we are talking here about things like how different groups greet each other, eat food, have relationships or celebrate important events.

But the more you think about this, the more you realise both how complicated all of this is, and how little we ever stop to think about it. All of these things are so habitual, so a part of the everyday that they feel 'natural' and we rarely pause to notice that they are not. There is no intrinsic reason why it is better to shake hands than it is to kiss or embrace when meeting, for example, or to eat curry and not cornflakes for breakfast. Yet we all have a sense of what the rules are and our place within them. If someone breaks one of those rules – by, for example, not showing respect to an elder, or by passing something with their left hand, or jumping the queue – it can cause offence or even outright conflict.

In fact, however, this goes beyond rules to something more fundamental about how we understand the world and our place in it. Culture in this sense is the story that we tell ourselves about ourselves, and the communities that we are a part of – and that story reflects the reality of that society, but also shapes that reality and in many senses *is* that reality. Our beliefs are constitutive of how we see and understand the world, and yet:

> Many of our beliefs exist in a semi-inchoate form that one might call the social unconscious – the vast repository of instincts, prejudices, pieties, sentiments, half-formed opinions and spontaneous assumptions which underpins our everyday activity, and which we rarely call into question. In fact, some of these assumptions run so deep that we probably could not query them without some momentous change in our way of life, one which might make them fully perceptible for the first time. (Eagleton, 2016, p49)

In other words, realising that so many of the things we do all of the time are not 'natural' but are in fact learned habits, often requires something very drastic to happen – for example, for us to leave the social group that we are used to by living abroad, or going to university, or both, and meeting people from other social groups and cultures that have a different set of habits and practices. If we are talking about the whole way of life of a group of people, we are not just talking about habits like how we eat or celebrate birthdays, but also about what we think is right and wrong, or good and bad – in other words, the values by which we live our lives, because a 'whole' way of life can be taken to mean a *good* way of life as well as an *entire* way of life. What constitutes a good relationship? A good friend? A good father? A good job? A good *life?* Or indeed, a good argument?

In this way, as Eagleton (2016) puts it, the word culture 'signifies the symbolic dimension of society as a whole, permeating it from one end to another ... There can be no distinctively human activity without signs and values' (p11). The desire to understand the world better, to produce and share knowledge, is one of the most distinctively human activities, and these signs and values provide the context within which we ask such questions, ascertain the facts that will help us to answer them, and build arguments about what those facts mean.

When thinking about academic study, the idea of 'culture' can often be used in a condescending way, as if it were a superficial consideration that a true academic could transcend and ignore when pursuing objective truth. But this is a simplification and a misunderstanding. As much as we might like to think otherwise, universities are a part of culture and society, and society is both 'an aggregate of millions or billions of individual choices' and also 'a complex, recursive dynamic in which choices are made within institutions and ideologies that change over time as these choices feed back into the structures that frame what we consider possible ... Which is just to say that we are not free to choose how we live any more than we are free to break the laws of physics. We choose from possible options, not ex nihilo' (Scranton, 2018).

The same is true of academic argument. Academic thought does not occur in a vacuum, and the range of possible options from which academic arguments are constructed and by which they are shaped include social and cultural assumptions and values. As we shall explore in a moment, academic disciplines in fact have their own cultures, which further complicates matters – but before we move on to that, let's consider the political and the economic.

Political and economic

Political and economic questions link directly to the questions of social and cultural values and in some ways are inseparable from them. Political questions are effectively a translation of what a culture thinks are the right ways of doing things into actual rules and governance. Economic frameworks provide a way for making decisions about what we spend money on based on these values and rules, and so on.

But there is more to it than this. There are lots of different ways of looking at politics and economics and these approaches are not just about the values themselves, but arguments about the best way to achieve and protect those values. So, electing a socialist government will lead to a different set of ideas being employed about what is the best way to achieve things than a capitalist government, for example. Both will claim to be interested in freedom, fairness and the protection of their citizens, but the former will tend to emphasise the collective, and represent the state as a body that should intervene heavily in the economy and people's lives to deliver the maximum good for the largest number of people, while the latter would focus on the importance of individual freedom and minimising the role of government.

It is easy to see not only how this leads to arguments in the day-to-day sense of conflicts over how countries should be run, but also arguments in the more academic sense where these positions drive research and further questions to try and prove and disprove positions and create new ways of thinking. Any question asked about the best way to govern or spend money will automatically take place in that pre-existing context, in which certain decisions have already been made, and any that have not will be made if not using, then at least in relation to, pre-determined value judgements.

The key here, again, is that any attempt to think about political and economic decisions does not happen in a vacuum, it happens in relation to existing positions – even if it is only to reject them. If you are completely against something, you are still defined by it, or at least your relationship to it. The idea of *WYSIATI* and *confirmation bias* also tells us that any decision we make or argument we build to advocate a particular position will inevitably be hugely influenced by the existing context.

No form of knowledge can think itself outside of these political and economic questions. Universities, wherever they are in the world, exist within local, national and international political and economic contexts, and the knowledge they produce inevitably has a relationship to those contexts.

Disciplinary frameworks

This is not to say that universities, or the knowledge that they produce, is reducible to, or controlled by political powers, be they local or global.

On the contrary, the university exists as a space to challenge and question, and to create the new. But again, this newness does not come from nowhere, but is forged out of the available possible options. As Bridget Adams (2009) puts it in her guide for students of psychology, 'even in the most scientific of physical sciences, cultural bias influences the kinds of research that will be undertaken and the ways in which funding is allocated' (p14). This is not just a question of which questions are explored and which receive funding, 'the dominant ideology in the Western world ... is white, male, middle-class capitalism. We can also add able-bodied and heterosexual. The majority of people in positions of power and influence in politics, business and higher education fit that description and their views on what is good, desirable and the right way to do things prevail. Everyone else is "other" and at a disadvantage' (p14).

In order to understand this claim, we can roughly translate the term 'ideology' as meaning something like what I have so far called 'theoretical frameworks'. We'll look in more detail at what this word means both later in this chapter and the next. For now, consider how a combination of confirmation bias, WYSIATI and the fact that the Western world and its university system remains predominantly white, middle-class and male affects the ways in which knowledge is constructed.

Even this does not tell us the full story. The university as we currently conceive it is a recent phenomenon (see Chapter 1), and many of the subjects that are studied within it are also relatively recent inventions. 'Social Sciences' such as sociology, economics or politics, as we understand them today, developed in the nineteenth century (Wallerstein, 1989), and the same is true of many subjects in the sciences, arts and humanities. As we explored in Chapter 1, the very idea that going to university is a good idea is born out of a whole range of assumptions, and the ways in which disciplines set themselves up as capable of producing knowledge about the world is the same. Psychology is considered a science, for example, but sociology is not. What does that mean?

Part of the answer to that question comes from understanding that disciplines (broadly, and in specific countries and institutions) have cultures of their own. Therefore, they have the same features as any other culture – they have their disciplinary unconscious, they have political and economic concerns, and they have their stories that they tell themselves about who and what they are. It is important to psychology to consider itself, and to be recognised as a science, for example, as this has connotations of seriousness, rationality and rigour that are important to maintaining not just the legitimacy of university psychology departments, but also an entire approach to understanding the way that human brains work and the apparatus that goes along with that – from psychiatrists and psychotherapists to mental institutions, and at what point people are no longer considered legally accountable for their actions or are considered too unwell (or too dangerous) to be allowed to be a part of society. In other words, the scientific status of psychology as a discipline has very real consequences – up to and including depriving people of their freedom.

Other, more prosaic examples of disciplinary or institutional contexts that might impact you as a student are related to, for example, how much critical thinking you are expected to demonstrate. Critical thinking is often held up as a key requirement for a university student – many books have been written on the subject – but this will mean different things in different subjects. In economics, for example, 'it's extremely important to be sure of how much critical thinking your lecturers want you to demonstrate ... Some lecturers encourage it, rewarding it highly when marking; but you should be aware that others want you to focus on explaining the models as they are rather than critiquing them. You need to get the balance right' (Mallard, 2012, p53). Different educational cultures will also have different perspectives on the level of criticality that is desired – or even permissible. 'Critical thinking', then, might be a common feature, but what that actually *means* in practice varies hugely.

Part of what is at stake here is *authority*. Think about this for a moment – what does it mean to you for something, or someone, to be an authority – that is someone who you can and should listen to, and someone you can trust? In an academic context, we talk about lecturers being experts and therefore we think that what they tell us can be trusted and believed. This is true in pretty much all educational contexts, but how that manifests itself might be very different. For example, when you were at school, were you expected to ask lots of questions and challenge your teachers, or were you expected to treat them with deference and respect, and learn what they told you and be able to repeat it back? Now scale that question up and think about who and what you are encouraged to question on a social or cultural level, and who and what you are not. Think about how you feel about public protests, for example – are they a good thing or a bad thing? How do you feel about the police or about politicians?

Even when we are not consciously aware of an argument being underpinned or authorised by an authority ('you should believe this because an expert says it is true'), there will be some kind of implicit appeal to authority. Often this is simply that we should accept an argument or idea because it has been arrived at objectively using the accepted methods of experiment and reason – as Toulmin (2003; see Chapter 2) would put it, this is the warrant that underpins how we can say that our evidence leads to our conclusion.

It's important to realise this not because it is wrong to trust knowledge gained in this way. The system of modern rational and scientific thought that I have very simplistically summed up here is the foundation of the modern world and the best that humanity can currently achieve. The reason that all of this is important is because it shows us that academic study and academic argument can never escape context, and that they are not immune from all of these considerations, no matter how much they would like to be.

Theoretical

The final type of framework that we need to consider is one that I shall call here 'theoretical', or that which cuts across all the categories that we have looked at so far. Patriarchy is one of these, and Bridget Adams (2009) identifies Marxism, feminism, multiculturalism and gay rights as 'opposing ideologies' to the dominant white, male, heterosexual ideology in a university context – which is in itself interesting. What we are talking about here is ideas that are cross-cutting and change all of the things that we've been discussing so far.

Some of these can be seen as identifying with, or being interested in a particular grouping or particular cause – in the case of those listed so far that would be women's rights, workers' rights or the interests of the non-white and non-heterosexual. It is interesting when you think about what proportion of the world is covered by those groups that these are often considered to be representing 'minorities'.

So, you can, for example, have a feminist view of society and culture, of political and economic factors or of the way in which academic study/work occurs in different disciplines (feminism is just the name for the belief that women should be treated equally to men). There are also cross-cutting ways of approaching knowledge – the main ones of interest to us here being structuralism, post-structuralism and deconstruction. At a very simple level, all three of these are concerned with not just what we say or know, but *how* we know it, looking at the relations between things, or in the case of deconstruction, 'the inadequacy of one's categories of analysis' (Bonnett, 2008, p15). Terry Eagleton and Raymond Williams, who I have quoted extensively in this section, are both Marxist cultural critics who, even if they would not accept the label, are clearly influenced by structuralism and post-structuralism, as is most of the way I have approached this book so far.

For our purposes here, we do not need to explore these complex concepts in much more detail – what is important is to know that all of these various different frameworks are often referred to as 'ideologies' – which is a word that has already been used here to cover a range of meanings. Before we look in more detail at what it means, though, let's think a little more about what all of this has to do with building arguments at university.

What is literature?

To explore this question, let us consider one discipline in particular – literature. The idea that literature is something that could be studied at university is in fact a twentieth-century invention and before we can ask ourselves the question 'what does it mean to *study* literature?' we first have to answer the question – what *is* literature?

This may not be a question that you have ever considered before, and it might be one that is far away from what you intend to study at university. 'Literature' also has many different meanings in different places. Whatever your background, however, it is likely that you have an idea of what literature is, even if you've never tried to put it into words before.

TASK 3.2

Make notes on your answer to the question – what is literature? Try and think about this in as many different ways as you can, for example by generating other questions such as:

- What examples of 'literature' can you think of? What sort of books and what sort of writing are, and are not, 'literature'?
- Who decides what is and isn't literature?
- Are there particular qualities that something has to have to be considered literature?
- When people study literature at university, what do they study?
- … and so on.

There are many possible answers to this question and all of the questions that it generates. You will certainly have been aware that some types of writing are considered literature, and that others are not. In most cultures, literature is highly prized, and many countries have key literary figures that are a vital part of their national identify – for example, Shakespeare, or Cervantes, or Tagore. When answering this question, you may have thought about the quality, clarity or beauty of the language, or the importance and interest of the subject matter. You may have thought about fiction and poetry as opposed to factual texts, the power of certain types of writing to inspire emotion, or to make people think about the world in a different way. You may have thought about literary awards, such as the Nobel prize, or of critics condemning popular culture as less worthy and *not* literature. All of these are common ways of thinking about how to define literature, and indeed for why it is (or isn't) worth studying.

Terry Eagleton, in his essay *What Is Literature?*, however, says that it is impossible to identify any set of common features in writing that is called literature, not least because those features vary from culture to culture and across time. In other words, different sorts of text have been considered literature at different points in history and in different places. Instead, he says:

> John M. Ellis has argued that the term 'literature' operates rather like the word 'weed': weeds are not particular kinds of plant, but just any kind of plant which for some reason or another a gardener does not want around. Perhaps 'literature' means something like the opposite: any kind of writing which for some reason or another somebody values highly. 'Literature' is in this sense a purely formal, empty sort of definition. (1996, p8)

At first glance, this might seem like an unsatisfactory answer – after all, if there are no rules to decide what is and isn't literature, other than that some people, somewhere, value it, then surely anything can be, and there is nothing coherent for us to study? Eagleton is emphatically not arguing this, however. Rather, his argument is about the very nature of labels such as 'literature':

> All of our descriptive statements move within an often invisible network of value-categories, and indeed without such categories we would have nothing to say to each other at all. It is not just as though we have something called factual knowledge which may then be distorted by particular interests and judgements, although this is certainly possible; it is also that without particular interests we would have no knowledge at all, because we would not see the point of bothering to get to know anything. Interests are constitutive of our knowledge, not merely prejudices which imperil it. The claim that knowledge should be 'value-free' is itself a value-judgement. (1996, p12)

For Eagleton, then, it is not that we should be trying to escape value judgements in order to arrive at objective description, but rather that we must recognise that value judgements are inherent to *all* knowledge. This is because:

> If it will not do to see literature as an 'objective', descriptive category, neither will it do to say that literature is just what people whimsically choose to call literature. For there is nothing at all whimsical about such kinds of value-judgement: they have their roots in deeper structures of belief which are as apparently unshakeable as the Empire State building … [It] is not only that literature does not exist in the sense that insects do, and that the value-judgements by which it is constituted are historically variable, but that these value-judgements themselves have a close relation to social ideologies. They refer in the end not simply to private taste, but to the assumptions by which certain social groups exercise and maintain power over others. (1996, p14)

In order to understand this claim, we need to know what Eagleton means by 'ideology'. This, he says is 'the ways in which what we say and believe connects with the power-structure and power-relations of the society we live in' (1996, p13). In other words, as we have already seen, what we say and believe does not come from nowhere, it comes from the society that we live in, and it is shaped by who/what holds the power in that society. What we privilege as 'literature' is not objectively, therefore, the best or most beautiful writing, as much as that may be what we think we are arguing about – or rather, if it is, it is what we think is the most beautiful or best because of the standards of judgement of our society, not because there is a neutral and eternal standard that we can compare it to.

This may seem like an argument that is applicable to literature, or perhaps the humanities more broadly, but one that has little use beyond that. However, this is not the case, and the same argument applies to all types of description, from the everyday all the way to the abstract and the academic. Think about the clothes that you are wearing right now, and the clothes that all the people around you are wearing. It is likely that you have quite strong feelings about what you would and wouldn't wear and that these will have changed over the course of your life. This is not just because your personal tastes happen to have changed – after all, it is likely that the people around you have also undergone similar shifts in their tastes. It is instead because of the society that you live in, and the power relations within that society, and how they shape your assumptions and everyone else's.

The same also applies just as much to the sciences or social sciences as it does to subjects in the humanities (or discussions about the type of clothes that are currently fashionable). Stephen Hawking, the famous physicist, argues that all scientific knowledge is based on theories, which are only models of the world and not a true picture of that world. He illustrates this with a discussion of how Einstein's general theory of relativity builds on Newton's theory of gravity, and shows it to be incorrect – but that despite this, 'we still use Newton's theory for all practical purposes' (Hawking, 1988, p10). Knowledge, he argues, is provisional, and its value is related to what we want to use it for and the context in which that judgement is made, not anything intrinsic to the knowledge itself. That means that the 'facts' that we arrive at are shaped by the context of the investigation that produced them – in other words, the question that was asked, or in Eagleton's terms, the 'particular interests' that were at play in asking that question.

In the next chapter, we will look in more detail at the term 'ideology', but for now let's consider one more definition. Louis Althusser (2001) says of ideology that it represents 'the imaginary relationship of individuals to their real conditions of existence' (2001, p109). This 'imaginary' does not mean that ideology is therefore to be valued less than some posited 'real'. Althusser is in fact talking about something closer to what Hawking is saying above, which is to say that ideology is a way of describing how we think of the world – it is the model that we have in our heads that we use to understand what is going on around us. This model is always to some extent a simplification – which is not necessarily a bad thing. As David Graeber and David Wengrow (2021), the anthropologist and archaeologist argue, 'one must simplify the world in order to discover something new about it. The problem comes when, long after the discovery has been made, people continue to simplify' (2021, p21).

For our purposes here, we don't need to follow this line of argument too much further. Rather, the insight we need to take from this is that if we are to be skilled in building and interrogating arguments, we need to understand both our own assumptions and the assumptions of others – that is, the theoretical or ideological frameworks that underpin our and others' positions. You cannot build good arguments without an understanding of ideology.

And there is nothing outside of ideology. There is no neutral, objective position from which to judge. We cannot go outside of the model of the world that we carry in our heads, we can only change it in the light of new information. This is not to dismiss the importance of objectivity or neutrality but rather to insist that the quest for an objective truth that can never be completely achieved is the engine behind all academic work and knowledge production.

In that sense, we can paraphrase Beckett as a maxim – 'Fail. Fail again. Fail better'. The knowledge that there is no final certainty we can arrive at is not a reason to give up, but the very reason why we do it all in the first place, knowing that there is always a better argument to come.

TASK 3.3

To apply this to your own studies, think about the following questions. You may not be able to answer some of them right now, but either way, it is worth keeping them in mind as you progress through university, as it can be an informative way to get an overview of how your discipline works and of your relationship to it.

- For the subject that you want to take at university or are considering taking, what is the object of study?
  - This may seem very obvious, but as with the case of literature, which we have looked at here, it is often more complicated than it first appears. Another way of thinking about this is to ask – what is the purpose of studying this subject? What are academics in this field trying to achieve? There may be more than one answer to this question, and the differences can in themselves be revealing.
- What other subjects are also studied as a part of learning about your discipline?
  - Academic areas often present themselves as being monolithic and completely separate but, in fact, almost all degree courses will involve a certain amount of cross-disciplinary study. For example, economics can draw upon sociology and psychology while politics might draw upon philosophy or marine biology on computer science.
- What are the main schools of thought (or theoretical frameworks, or ideologies ...) relevant to your subject?
  - This, in particular, may be a question that you cannot answer yet, but keep thinking about it as you go through your academic career, as it is vitally important to understanding the arguments that you will come across. One clue can be to look out for words ending in 'ism' or 'ist' – whether that be constructivist, positivist, feminist or communist. Other frameworks relate to particular theorists (for example, Marxism or Keynesianism) or different methodologies (for example, qualitative, quantitative or autoethnographic).

(Continued)

- Which of these schools of thought do you find most appealing and why?
    - Again, this may be a question that you cannot answer yet, but it is worth thinking about as you progress through your studies, in order to think about what appeals to you and why. For example, when I was an undergraduate, I was very drawn to both post-structuralism and postmodernism, partly because they seemed to put into words things that I already felt about the world and partly because they seemed very current and relevant. You will have your own reasons, but it is worth thinking about why one area of study, or idea, or approach particularly speaks to you.

Take notes on the questions below:

- What is the object of study for your chosen subject? What is the purpose of studying it?
- What other subjects are also covered as part of studying that discipline, either in general or on your particular course?
- What are the main theoretical frameworks relevant to your discipline?
- Which of these do you find most appealing?
- Why?

Summary

- We cannot understand arguments, or build good arguments, without thinking about the frameworks within which these arguments are constructed.
- Social, cultural, political, economic and theoretical factors all influence the arguments that we make and the knowledge that we produce. There are a number of ways of referring to these factors, including the term 'ideology'.
- Because of this, arguments and knowledge have a relationship to power and authority.
- Universities and academic departments are not outside of this, even if part of their function is to criticise and challenge dominant ideologies. Academic disciplines in fact have their own cultures and power relations.
- Objectivity is the aim of all academic work, but complete objectivity is not possible. It is important to remember this not in order to avoid needing to try and be objective, but rather in order to be constantly aware of the limitations of your own position and the tendency to think that you can transcend them.

FURTHER READING

Chapter 9: Understanding cognitive bias from Chatfield, Tom, *Critical Thinking* (Sage, London, 2018).
Huff, Darrell, *How to Lie With Statistics* (Penguin, London, 1991).
Kahneman, Daniel, *Thinking Fast and Slow* (Penguin, London, 2012).

NOTE

1. See, for example, the physicist Albert Michelson's claim in the early twentieth century that the 'edifice of science' (and therefore human knowledge) was nearly complete, needing only a 'few turrets and pinnacles to be added, a few roof bosses to be carved' (Bradshaw, 2001, p121).

4

Reading for argument

As with so much else in university study, the most important way to learn about arguments is through reading. It is not always immediately obvious, but while the content of the texts you read is vital to your studies, what you read also has an enormous amount else to teach you. When you are reading, you should think not just like a student, but like a writer, and consider what you can learn from the text you are dealing with, on every level.

For example, reading the texts that make up your discipline will tell you a lot about what is considered a good writing 'style' in your subject, what different genres of writing are used (see Chapter 2), how those genres are structured, and of course not only what arguments are of interest in your field but, just as importantly, what constitutes a good argument in the first place.

This is especially important, as a key complaint from lecturers across many different disciplines is that students are good at reading for facts, but not good at reading for argument.

In this chapter, we will look at how to pick out arguments in academic texts, both in terms of how to unpack complex texts, but also in terms of how to read for argument, and not simply for content. A key concept here is the idea of **active reading**.

Active reading

Everything that we discuss here about being an *active* and *critical reader* applies equally to being an *active listener*. At university, you will encounter as many arguments when listening as you will when reading, whether that be in lectures, in discussions with other students, or when watching videos. The strategies and questions that you can apply to reading a text are the same ones as you should apply when you are listening.

It is easy to think about reading as a passive process, whereby you, the reader, simply absorb what is there in front of you. The text is a source of knowledge, and you are an empty container, ready to be filled up with that knowledge.

While this metaphor is in some ways appealing, it is not a very useful one, especially for university study. In Chapter 1, I asked you to consider where the meaning of this book comes from: is it from me, the author; from you, the reader; or from the text itself, that is the words on their own, separate from either of us, and what they mean in themselves?

In the metaphor above that sees you as an empty container, I am the source of meaning, which is then transferred to you via the text without anything being lost or changed along the way. The text is simply a means of transferring the message that I want to impart to you, the reader.

This is obviously problematic. As we explored in the first three chapters, our answers to any question, and our encounters with any text, are shaped by a huge range of factors that include, but are not limited to, our own backgrounds, our purpose in asking a question or reading a text, and the situation in which that asking or reading occurs. The meaning of a text is not stable or fixed therefore, but is rather created in each act of reading, by the specific reader and the context in which the reading takes place.

This might sound like an abstract and confusing way to think of it, but it can just as easily be seen as a liberating approach. Instead of meaning being something that is created by, and belongs to, the expert author, who bestows it on you, the reader, meaning is in fact something that is created by the interaction of the author, the reader and the text.

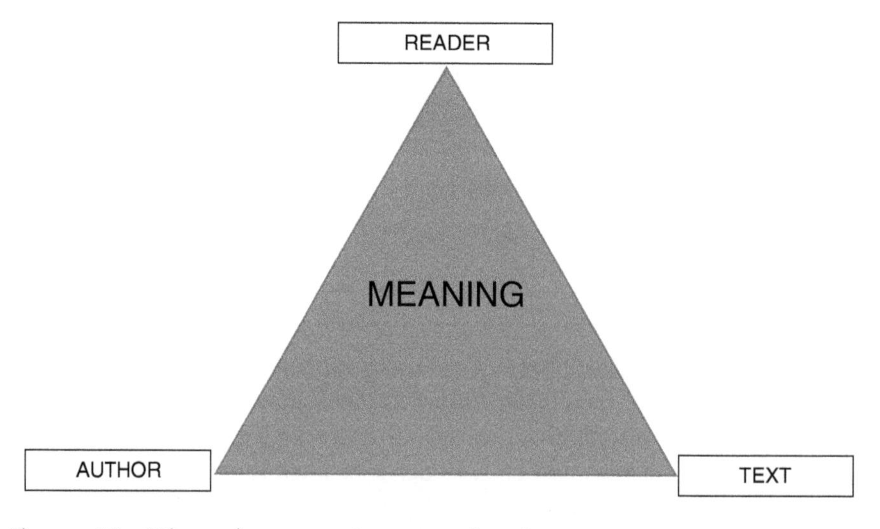

Figure 4.1 Where does meaning come from?

Rather than it being the case that there is one perfect reading – One Right Answer – which you get more or less correct (and, especially with difficult texts at university, usually less), the 'correctness' of the reading is determined by what it is that you want to get out of the reading at that particular point as much as it is by the content of the text. That is not to say that there are no wrong answers or that you can say anything you like. You can only base your answer on what is in the text, but effective reading for university does not have to involve a complete understanding of every aspect of the text and instead involves reading strategically and for a particular purpose (e.g. to write an essay, understand a lecture … or win an argument).

What that also means is that there is not a lack or a hole in you that the text is there to fill. Rather, you, the student, are the active agent here, the one with something to achieve, which the text is there to help you with.

This is one way to think about active reading. In this model, it is your activity that creates the meaning – you are the most important element. That might sound daunting or confusing but it is in fact very practical. This will not only ensure that you get the most out of your reading, but can also help with the anxiety and stress that goes hand in hand with the high volume of reading expected of you at university.

Let's see what this looks like in practice. The first thing to note is that this sort of approach relates to the idea of *critical thinking*. As discussed in Chapter 1, being critical at university is about questioning, about analysing and interrogating ideas rather than simply accepting them. This is one key element of active reading – to not approach a text passively, but to approach it with questions and strategies to make sure that you are not just accepting what is presented to you, but are engaging with it in the most effective way to improve your own understanding and achieve your own goals.

These can include a range of different types of question, but one useful way to think about this can be approaching it via the three sides of the triangle above – namely the text, the author and the reader.

For example:

Details of the author/text:

- Who wrote this text? Why?
- When was it published? Is it up to date?
- What sort of text is it (e.g. an academic essay, a book, a webpage, an encyclopaedia entry)?
- How does it relate to other texts on this topic?

Evaluation of the text:

- Is it of a suitable quality?
- Is it internally consistent?
- Are points well supported by evidence?
- Is anything missing?
- Is the author writing from a particular standpoint, and are they drawing on a particular theoretical framework?
- What is it trying to demonstrate/what question is it trying to answer? Does it succeed?

As well as these general sorts of questions, you should also be asking yourself practical questions related to the third side of the triangle, namely you, the reader. In particular, you should be thinking about your position as a student. So, for example:

- What is your purpose in reading this text?
- What do you expect to get out of it?
    - Are you reading to prepare for a weekly class, to write an essay, or for background research? Or just for general interest/information?
- Were you told to read this text, or did you find it yourself?
- Is this a topic you are familiar with or one that is completely new to you?

It is also worth returning to the questions that we discussed in Chapter 2, especially if the text is something that you've been told to read by a lecturer – namely:

- Why has my lecturer asked me to read this?
- What are they expecting me to get from it?
- How does it fit in with the rest of this module and the rest of my degree?

We will discuss this in more detail in a moment and think more about strategies that we can use to become better, more active readers, but for now let's practise by considering a non-academic text.

Memes and fake news

Over the past few years, the question of how we know what information we can trust has become ever more pressing, not least because of the proliferation of, and interest in, 'fake news'. The internet gives us access to ever more sources of information, and claims that attempt to convince us of competing views of the world. But how do we know which to believe?

Consider the following. In the run up to the 2016 US presidential election, a huge amount of material was produced and shared, via both traditional and social media. Much of this was in the form of 'memes', or short, often image-based texts that are designed to be easily and widely proliferated.

Many of these concerned the businessman and reality TV star turned presidential candidate, Donald Trump. In the US there are two competing political parties, the Republicans and the Democrats. Trump was the Republican candidate.

TASK 4.1

Look at the following example of a meme and decide – is this true? For the moment, don't think about whether you agree with what the text says – literally just decide, did Donald Trump really say this? List reasons why you think it might be true that he did and anything that suggests to you that it might be false.

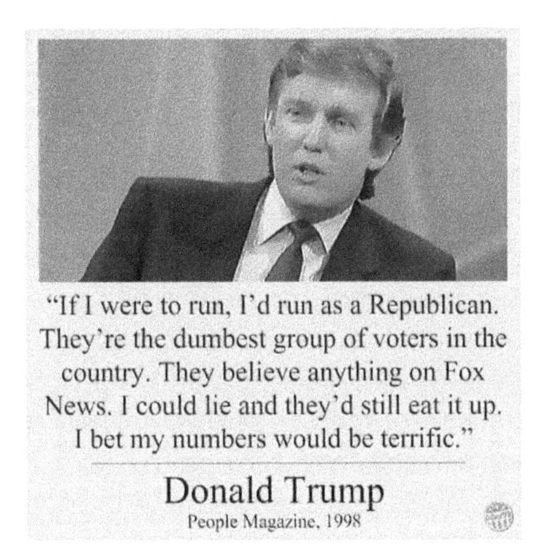

Figure 4.2 Did Trump really say this?

Remember to try and base your conclusion only on the evidence that is contained within the meme itself.

There are lots of possible reasons that you might have thought that this was true if you came across it online in 2016. The first would have been the thing that makes all fake news so effective – the knowledge that it has been shared in the first place. The very fact that this is circulating widely suggests that other people believe it, and makes it easy to think that someone, somewhere must therefore have checked whether or not it is true – otherwise surely it wouldn't have been passed on?

There are also lots of details within it that make it seem plausible. It sounds like Donald Trump – anyone who is familiar with how he talks will recognise his rhetorical style. It also sounds like the sort of thing he might say – and indeed is very similar to other proclamations he has made, such as that, 'I could stand in the middle of Fifth Avenue and shoot somebody and I wouldn't lose voters'. It also has a reference that seems to tell us where this quote comes from – i.e. *People Magazine*, 1998 – which suggests that it is *verifiable* (even if we don't go to the effort of verifying it ourselves), and the fact that there is a picture of Trump above also somehow adds verisimilitude.

However, there are also many things here that should make us suspicious. Firstly, the picture is odd. This is reportedly a quote from a magazine interview and yet the photo seems to be a random still from a TV appearance, and one in which Trump looks a lot younger than he did in 1998. The reference is also a lot less reassuring than it seems, because it is incomplete. *People Magazine* is weekly, and the reference does not give us a date or a page number, and so to verify this quote we would need to look through an awful lot of pages to check.

We also need to consider two of those example, active reading questions from above – *who wrote this and why*? If you look at the bottom right-hand corner of the image, you will see a little badge that says 'the Other 98'. This is the group that made and circulated the meme, and they are a left-wing group that describe themselves as dedicated to resisting the two per cent of billionaires and big business that control so much of the world's resources. It seems unlikely that their purpose, therefore, is positive towards the billionaire, ultra right-wing Donald Trump. That does not mean that the meme is fabricated, but it should make us cautious.

We will come back to the question of 'why' this was produced in a moment, but there are a couple of additional points that we need to consider before deciding whether or not this is true. Firstly, there is the notion of *confirmation bias* (see Chapter 3) – or the fact that in this case, many of the people who encountered this meme would have wanted it to be true because of their antipathy to Trump and so would have been primed to believe it.

There are also some clear *factual errors* here. The most important of these is that, while by the time of the 2016 election Fox News was a famously right-wing and Trump-supporting media outlet, in 1998 it was neither famous, nor particularly known as being right-wing. It did not achieve this status until after George Bush's election in 2000, 9/11 in 2001 and the Iraq War of 2003. In other words, this quote was written with 2016 in mind, not 1998 (see LaCapria, 2015).

As you may have guessed from the above, the quote is not real – Trump never said this. Even without knowing some of the factual points above, it is largely possible to work this out just from looking at the meme. Why, then, did someone create and share it? What was their purpose – or to put it another way, what argument were they trying to make?

Take a few moments to think about why you think this image was created and shared.

It is of course impossible to say exactly what the purpose of this image was – or even if it was actually made by the Other 98. However, there are several arguments that it could be seen to be making.

Firstly, the meme uses a credible impersonation of Trump to remind readers how many disparaging things he has said about people and to make him seem less likeable. It also suggests that, far from being on the side of the Republicans he is claiming to represent, he actually holds them in contempt.

Overall, then, the argument that would seem to be being made here is – do not vote for this man, either because he is just as much of a liar as you always thought he was (if you are a Democrat), or because he is actively laughing at you, the Republican, he is trying to convince to vote for him.

This example is not directly transferable to an academic context, as at university you will (hopefully) not come across many examples of texts that are actively trying to lie to or trick you. However, it is an example of how active reading can help us to evaluate a source, and also, crucially, think about what the implicit purpose of a text is – what argument is being made, in other words, and what that argument tells you about what you are looking at.

It is also worth noting that the way in which I presented the various possible factors that might signal that this quote was real or fake is also an example of structuring an argument for effect. We will return to this in Chapter 6.

Reading complex text

Before moving on to a more holistic example of reading for argument in an academic context, we need to consider another side of active reading, which is to think about how you can understand dense and complex text just through reading actively and attentively.

This time, let's use a paragraph from an academic journal article. It is taken out of context and because of that, at first glance, is likely to appear quite complicated and difficult to understand.

However, by considering what we can work out just from the paragraph itself, we will see how a critical and active approach to reading allows you to understand even difficult and unfamiliar ideas, without needing anything more than the text in question.

TASK 4.2

Read the following paragraph. What questions do you need to answer in order to understand what it is saying? What can you work out just from looking at the paragraph itself?

> We do not usually think of ideologies as institutions. But this is in fact an error. An ideology is more than a *Weltanschauung*. Obviously, at all times and places, there have existed one or several *Weltanschauungen* which have determined how people interpreted their world. Obviously, people always constructed reality through common eyeglasses that have been historically manufactured. An ideology is such a *Weltanschauung*, but it is one of a very special kind. It is one that has been consciously and collectively formulated with conscious political objectives. Using this definition of ideology, it follows that this particular brand of *Weltanschauung* could be constructed only in a situation in which public discourse accepted the normality of change. One needs to formulate an ideology consciously only if one believes that change is normal and that therefore it is useful to formulate conscious middle-run political objectives.
>
> Three such ideologies were developed in the nineteenth century – conservatism, liberalism, and Marxism. They were all world-systemic ideologies. (Wallerstein, 1989, pp44–5)

There are lots of different aspects that we could choose to focus on here, but two key terms that we clearly need to understand in order to work out what this paragraph is saying are 'ideology' and '*Weltanschauung*'.

What can we work out about these two terms just from the paragraph itself?

Firstly, is 'ideology' a term with a simple or only one definition? No – otherwise this paragraph would not be necessary. Also, the fact that Wallerstein says 'using this definition of ...' implies that there are other definitions that could have been used – some of which we have already looked at in Chapter 3.

What is the relationship between the two terms? Clearly, *Weltanschauung* comes first – as an ideology is a kind of *Weltanschauung*. Ideology thus seems like a more complicated concept than *Weltanschauung*, although one that is related to it.

So what does this German word *Weltanschauung* mean? Is there anything in the text that can help us work this out? Yes – the sentence 'people always constructed reality through common eyeglasses that have been historically manufactured' is giving us a definition here, even though it is not explicitly stated. It's a complicated metaphor in some ways, but also a fairly clear one. It tells us that a *Weltanschauung* is a way of looking at the world (a set of 'eyeglasses'), and one that changes over time (is 'historically manufactured') and doesn't just look at the world but also shapes it ('construct[s] reality'). So a *Weltanschauung* is a way of understanding and describing the world.

What else from the text can help us to understand what 'ideology' is here?

It's 'consciously ... formulated' and has 'conscious political objectives' – so it is something that is deliberately made and is something that is related to politics. It also wants to achieve something – it is not just a way of looking at the world, but a way of changing it. Ideology, here, is also related to the idea that 'change is normal' and that such change can be deliberately brought about and controlled.

The paragraph also tells us that conservatisim, liberalism and Marxism are all types of ideology.

Are these words already familiar to you? If so, what do you know about them and does this help you?

Even if you don't, the meaning of the words 'conservative' and 'liberal' are hinted at by their everyday meanings, and even the briefest investigation would tell us that these were names for types of political approach.

At this stage, we might need to go beyond the text to fully understand it. If you were reading this as part of your studies, you would probably want to check the meaning of 'ideology' anyway. Here's one dictionary definition (taken from a bog-standard, non-academic, first-page Google search):

1. A system of ideas and ideals, especially one which forms the basis of economic or political theory and policy.
2. The set of beliefs characteristic of a social group or individual.
3. The science of ideas; the study of their origin and nature.

Which of these definitions is Wallerstein using?

(1)

Based on all this, what do you think *Weltanschauung* means? Does it correspond to any of the other definitions?

(Yes – 2)

Here is a definition of *Weltanschauung*, taken from a similar source:

- 'a particular philosophy or view of life; the world view of an individual or group'.

Having now used largely only what is already there in the paragraph, we have come to a pretty good understanding of the definition of ideology that is being used here – it is a worldview that is used to decide the ways in which political change should be achieved, and what that change should be.

Now, let's use that to try and put together an understanding of what Wallerstein's argument is here – that is, what is he saying about ideology?

TASK 4.3

Take notes on what you think Wallerstein's argument is here, in preparation for writing a short paragraph of your own, explaining his main point.

Now *write a short paragraph explaining this paragraph from Wallerstein* using a mixture of your own words and short quotes (i.e. less than 10 words):

Example: 'The term 'ideology' can be defined in many different ways, but here Wallerstein (1989) is using it to refer to a set of political ideals with a clear aim in mind, rather than simply a '*Weltanschauung*' or worldview. For Wallerstein, such ideologies could only develop when 'public discourse accepted the normality of change' (p44), and it therefore became necessary to think about medium- and long-term political aims. The three that emerged in the nineteenth century were conservatism, liberalism and Marxism.'

Strategies for active reading

Having looked at how we can use a critical and active approach to understand a text's argument, let's look at a slightly longer piece and consider ways that we can usefully read for argument.

Before we do that, let's consider some of the other strategies that we can use to make sure we are reading actively and critically. We've already discussed some questions that we can use to think about and evaluate the text (see pp.65–6), and our own aims in reading, but now, let's think about the whole process in some more detail.

1. Find your context

In other words, where are you coming from? Why are you reading this text and what do you want to get out of it? Who is the author and what question are they trying to answer? This links back to the trigger questions from Chapter 3, and also to the questions that we looked at earlier in this chapter. Think, for example:

- Why am I reading this? Why did my lecturer set this reading/why did I choose this text? Are there any specific questions I want it to answer?
- Who wrote this and why? What sort of text is it?
- Does it relate to any debates or theoretical frameworks?
- How does it link to my studies, either in terms of this module or my whole course?
- Does it link to anything else that I've already read?

... etc.

This will help you to think about your purpose for reading, and what exactly it is that you want to get out of a text (or lecture, or discussion).

If you want to, it can be helpful to set specific questions that you want a text to answer or information that you expect it to provide – for example, linked to a particular assignment you are working on or a topic you are studying.

2. Ask questions

As you read, ask questions of the text. Think, for example:

- What is the main argument here?
- What evidence or claims back up that argument?
- Is there anything missing or not considered? Why were certain examples chosen and not others?
- Are other arguments used here? If so, what are they?
- Is the text answering the questions I want it to? If not, why not?

... etc.

Keeping questions like this in the front of your mind will help to make sure that you are not passively accepting what the text says but remaining critical. It will also help to provide you with a focus for engaging with the text, and recording that engagement.

3. Take notes

Don't just highlight or underline text as you read or copy down what is being said in a lecture. Just marking a passage as interesting by underlining or highlighting it is *passive* whereas writing down *in your own words* what point a text is making is *active*, and will help you to engage with and understand what you are reading or listening to.

This is a very simple, but very important point. One of the best ways to be an active reader or listener is to *take good notes*. These will help you massively later, when it comes to revising or writing assignments, but even if you never look at them again, the act of writing them in the first place will ensure that you properly learn something and remember it.

Make sure to keep good records of what your notes are based on, including page numbers for where specific points are taken from, and be particularly clear about where you have written anything down word-for-word, exactly as it appears in the text. This will help you to draw on the notes easily later and to avoid plagiarism.

4. If you don't understand, keep reading

It can be very tempting to stop every time you come across a word you don't understand or if something seems too difficult. However, it is often the case that, as with the Wallerstein paragraph above, the context can explain it for you. So keep reading, look for clues and see if you can either work out what is being said or enough about what is being said for your purpose at that moment.

If it's still not clear, you will need to stop and work out why you can't understand something – don't simply keep pushing on if you really are not sure what a text is about. But it's disruptive to keep stopping and starting, and breaking your attention in this way makes it harder to understand and retain information. As with the Wallerstein exercise, often a text will give us everything we need to unpack it, and all that we need to do is to consider more carefully what is happening.

5. Reflect afterwards

The act of reading continues after you stop looking at the text. Continuing the conversation after your reading (as Ramage et al. 2001, p25 put it) will help you to both check your own understanding and fix that understanding in your mind. There are a number of ways you can do this, with the most simple being just reviewing and tidying up your notes. But it is also really important to follow up on the questions that you set yourself before or during your reading and/or consider some new ones. For example, after reading, you can see if you can answer questions such as:

- Do I understand why my lecturer set this text?
- Did I get what I wanted to from reading this?
- The most important argument in this text is ...
- What do I agree with here? What do I disagree with or have doubts about?
- Does this relate to any other arguments I've encountered?
- Did this text tell me what I expected? Why/why not?
- Was this text useful for me? Why/why not?

There are many more of these, a lot of which will be generated by the particular text or argument in question. These are just some examples to get you started.

Reading for argument

Now, let's go through this whole process using an example text. As we are not studying a particular module, we cannot interrogate our context as we usually would here. Instead, let's create our own context, based on the text that I am going to ask you to read.

TASK 4.4

Consider the following ethical dilemma. If you found out a shop made its goods in factories where conditions were poor and dangerous for workers (these factories are often referred to as 'sweat shops'), would you stop buying things there? What arguments can you think of that you should, or shouldn't?

Take notes and think about what your answer is.

At its simplest, this is a yes or no question – should we buy products that are produced in poor conditions? – but it's unlikely that you'll be 100 per cent either way. Remember the idea of setting yourself a scale from Chapter 1, and see where you would put yourself on the line.

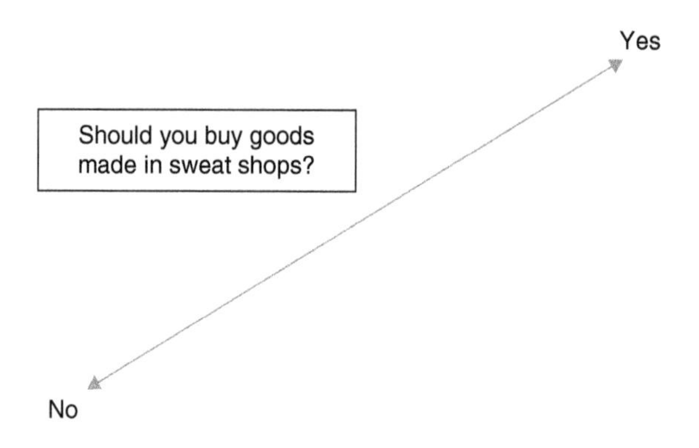

Figure 4.3 An ethical dilemma

Having thought about the general context, let's consider the text itself. It is a blog piece called 'Choosing Well' by the sociologist Lynne Pettinger (2012), taken from her blog *No Way to Make a Living*, and was written shortly after a fire killed 21 workers in a factory in Bangladesh in 2010. This was one of several such incidents that drew a lot of attention to the fact that clothes sold by successful brands around the world were produced in dangerous and unsafe conditions by poorly paid workers.

As we have explored above, a key aspect of active reading is to evaluate the text before we read and to see what we expect from it. There are lots of things to consider so let's see if some of the questions we discussed above can help us.

See what the brief amount of information I have given you can tell you by considering the following questions and taking notes:

- Who wrote this?
- What sort of text is it?
- Why was it written and how does it relate to other texts on the topic?
- What does this tell us?

Who wrote this?

This is written by sociologist Dr Lynne Pettinger. She is a university lecturer and published academic.

What sort of text is it?

This is a blog piece, written for a website called 'No Way to Make a Living', which was described as 'a sociological space for exploring what work is like ... in today's world' (Pettinger and Lyon, 2012).

Why was it written and how does it relate to other texts on the topic?

A website/blog provides the opportunity to discuss current, relevant questions in a more informal manner than academic formats (such as journal articles). Pettinger describes the site as 'a collaborative project' that emerged from 'frustrations with ... the forms of knowledge we can convey in academic publishing' (Pettinger and Lyon, 2012). From this it is clear that while a piece such as this allows some freedom from academic norms and constraints, the aim is to pursue the same questions and with the same level of depth and complexity. Therefore, we can expect the text to be a mixture of personal and academic reflections. We can expect it to be more journalistic in tone than a 'normal' piece of academic writing but still to be backed up and informed by academic research.

As Pettinger is a sociologist, we can also expect that the piece will relate personal and individual concerns (e.g. should I buy clothes from a brand that uses sweat shops to manufacture its clothes?) to broader social contexts, systems and issues.

Now, before we think too much about the specific argument, let's think about the second set of questions that were introduced at the start of this chapter – that is, those that were aimed at evaluating the text.

Read through the text below, and consider the following questions:

- Is it of a suitable quality for academic use?
- Is it internally consistent?
- Are points well supported by evidence?
- Is anything missing?
- Is the author writing from a particular standpoint and are they drawing on a particular theoretical framework?
- What is it trying to demonstrate/what question is it trying to answer? Does it succeed?

Choosing well – Lynne Pettinger

24 March 2010 – blog post from 'nowaytomakealiving.net'

H&M, the Scandinavian fast-fashion brand, has just opened a store in the town I live in. It opened a few days after a fire killed 21 employees of a knitwear factory in Bangladesh which is subcontracted by H&M to make those cute stripy jumpers, and that really useful little black cardy.

My friend called me last Saturday, 'let's meet in Hennes', she said. I agreed. I thought I'd just have a look and not say anything to her. But I couldn't help myself (story of my life).

'I'm not buying anything here, after all those people died'.

That made it impossible for my friend to even try anything on (I think she might go back without me; and I will confess to her now I was wearing something I'd bought in H&M last year the next time we met).

I've read Naila Kabeer's (2000) *The Power to Choose* and was persuaded so well by her arguments against reading Bangladeshi working women as cultural dopes, stepping blindly into exploitative paid work whilst carrying the burden of housework and facing down challenges to their reputations as good women. Kabeer's incorporation of how culture is 'woven into the content of desire itself' (2000: 328) is persuasive. Women chose paid work outside the home and still counted as good, they liked working in a clean place for good wages far more than labouring in a field, and took pleasure in contributing to meeting their family's desire for more income.

Kabeer gives the garment workers agency and voice. They are not an innately malleable, grateful, reserve army of nimble fingered knitters; they are not victims of a disorganized capitalism where feminism and neoliberalism combine to turn 'a sow's ear into a silk purse by elaborating a new romance of female advancement and gender justice' (Fraser, 2009). For Fraser, the normalisation of the dual income family working for low wages in insecure employment marks a failure of feminism, for (without realising it) privileging choice no matter what.

When War on Want describe sickening factory conditions and I read of these injuries and deaths, this is damage, and Fraser's is the line that persuades me. As I don't want my consumption practices to cause harm, that means no to H&M. In turn that means job losses, either because political pressure on H&M makes them choose a new subcontractor (one less famous for its working conditions), or because of the fall in demand caused by my bleeding, liberal, western heart. This is damage too. I'm not adding much to an unanswerable debate other than easing my own conscience by playing out the tensions: strong conclusions are impossible when there's only a choice between forms of damage.

References

Fraser, N. (2009) 'Feminism, capitalism and the cunning of history', *New Left Review*, 56.

Kabeer, N. (2000) *The Power to Choose. Bangladeshi Women and Labour Market Decisions in London and Dhaka*. London: Verso Publishing.

Answers

Here are my answers:

- Is it of a suitable quality for academic use?
  - *Yes, you could use this source as a university student. It is written by an academic and demonstrates complexity and nuance in thinking about the topic.*
- Is it internally consistent?
  - *Yes, although it can be a little bit confusing trying to pick out the overall argument, as the piece is so brief and refers to complex ideas only in passing. But the conclusion follows logically from the points made.*
- Are points well supported by evidence?
  - *Yes, Pettinger refers to two academic sources, although explanations are brief.*
- Is anything missing?
  - *The piece is very brief, but it consciously aims to consider both sides of a particular debate. It does so very briefly, but that is as much about the nature of the genre as any fault with the text itself.*
- Is the author writing from a particular standpoint and are they drawing on a particular theoretical framework?
  - *Yes – as well as being from a sociological perspective, the text also mentions feminism, and its central concern is with a feminist issue. In order to fully understand the argument made, it is important to consider this.*

- What is it trying to demonstrate/what question is it trying to answer? Does it succeed?
  - *It is trying to answer the question: should we buy products from brands such as H&M, even though we know that they are only able to sell these products so cheaply because they are produced in exploitative conditions (i.e. poorly paid and unsafe)? It does succeed in answering this – although the final conclusion is ambivalent, as we will see below.*

Having considered what we expect from the text, and given it a quick read to evaluate its quality, let's consider the argument it is making. As we do so, remember to consider steps 2, 3 and 4 from the list above – that is, ask questions as you read, take notes, and if you have any difficulty understanding what Pettinger is arguing, see if you can work it out from the context and keep reading for clues.

TASK 4.5

Re-read Pettinger's text above. Consider the following questions, and any others that you think are relevant:

- What is Pettinger's overall argument?
- What ideas/claims does she consider and use to back up her argument?
- What do you agree/disagree with?

Having read the blog piece again and taken notes, let's unpack the argument a little by following up with some more targeted questions:

- Where does the opening narrative change to analysis?
- How does the opening narrative relate to the main point that Pettinger wants to make?
- Are other arguments used here? If so, what are they?
  - What, according to Pettinger, is Kabeer's position on Bangladeshi women workers?
  - What is Fraser's stance?
  - Whose view does Pettinger most agree with – Fraser or Kabeer?
- What is Pettinger's overall argument?

In order to answer these questions, let's work through some of the more complicated parts of the text. Pettinger's blog post opens with a personal narrative that situates us in the ethical dilemma we've already considered – should we buy clothes from a shop when we know that they are made in poor conditions?

The text switches to analysis at the following point:

> I've read Naila Kabeer's (2000) *The Power to Choose*, and *was persuaded so well* by her arguments *against* reading Bangladeshi working women as *cultural dopes* ... (emphasis added)

This is where Pettinger switches from telling us an anecdote about her personal experience to analysing what this experience can tell us about the wider context and issues. This is signalled by the fact that she starts to refer to the arguments of others – i.e. to academic texts. The opening narrative serves to give us an individual, and therefore accessible, context to help us consider and understand the wider sociological considerations that these academic texts will help to illuminate.

What does this opening sentence of the analysis tell us?

- We know that Pettinger thinks Kabeer's argument is good as she says she 'was persuaded so well'.

But what is Kabeer's argument?

- Kabeer argues, according to Pettinger, that Bangladeshi women working in factories are not 'cultural dopes'. What does that mean? Let's carry on reading:

> ... as cultural dopes stepping blindly into *exploitative paid work* while carrying the burden of housework and *facing down challenges to their reputations as good women* ... Women *chose* paid work outside the home and *still counted as good, they liked working* in a clean place for good wages far more than labouring in a field, and *took pleasure in contributing* to meeting their family's desire for more *income*. (emphasis added)
>
> Kabeer gives the women *agency and voice* ... (emphasis added)

What does it mean to give people 'agency and voice'? Both of these words are likely to be familiar to you, although perhaps not as they are being used here. From the context of the previous paragraph, we can see that Kabeer is saying that these women are making their own choices, choices that are giving them power and positive outcomes, despite the fact that the work is 'exploitative' and culturally problematic ('challenges ... their reptuations as good women'). In other words, work in this context can be seen as giving women the right to speak for themselves ('voice') and as giving them the ability to make their own choices ('agency').

Let's keep reading:

> ... they are *not victims of* a disorganized *capitalism* where feminism and neoliberalism combine to turn 'a sow's ear into a silk purse by elaborating a new romance of female advancement and gender justice' (Fraser, 2009). For Fraser, the *normalisation* of the dual income family working for low wages in insecure employment marks a *failure of feminism*, for (without realising it) *privileging* choice no matter what. (emphasis added)

This next section reinforces what we have worked out above – Kabeer is arguing that these women should not just be seen as victims, as that contradicts the idea of them having 'agency and voice'. They are active participants in this working relationship, and not just 'dopes' who are being taken advantage of. But how does Fraser's point, somewhat abruptly turned to here, fit in?

Two key words here are 'normalisation' and 'privileging'. The complicated quote from Fraser is perhaps intimidating, but to understand what is happening here, you don't actually have to engage with that quote very much at all. 'Normalisation' we can work out from knowing what the word normal means, i.e. 'normalisation' means to make normal, while 'privileging' we can deduce from the context – it means placing value on.

Fraser is making the opposite argument to Kabeer, in other words, and saying that to cast Bangladeshi women as empowered because they worked in a factory is to 'normalise' low wages and bad conditions because of a misguided idea that a household having both parents working ('the dual income family') is automatically desirable and that having choice is always good, even if that choice is between 'low wages' and 'insecure employment', or labouring in a field and unpaid in the home. This is a 'failure of feminism' because it permits exploitation of women by sticking to the idea that choice is a good thing, no matter what, and assuming that an alternative to an oppressive position is automatically good simply because it matches a capitalist ideal.

As discussed in Chapter 3, the mention of this theoretical approach should make us think. What does this tell us about what we are reading? Feminism is concerned with gender equality, and in this case we can see that while the topic of this article is poor working conditions and what our relationship as consumers is with those conditions, it is also very explicitly focused throughout on the fact that it is *women* who are involved in this exploitative work in poor conditions. This has not been explicitly stated anywhere in the text, but as soon as you stop to reconsider it, it is clear that this is a vital element in understanding the argument that Pettinger is making. Everything that is being considered here is through the lens of feminism – that is, how this situation affects women and gender inequality, and the argument only makes sense through that lens.

Thinking of it in this way enables us to unpack the argument much more clearly.

The thread of the argument goes something like this:

Shops such as H&M sell cheap clothes, which they can only produce so cheaply by paying workers in Bangladesh poor wages to work in factories that are unsafe. The workers in these factories are almost all women. Therefore this is a feminist issue.

How does this help us answer the implicit question that underpins the whole text – should we buy products that are produced in such ways? The immediate answer would seem to be no as by doing so we would be participating in the exploitation of these women.

However, Pettinger finds Kabeer's argument that that these women are not (just) being exploited, and that their jobs as such have given them power, freedom and agency in ways they would otherwise be denied, persuasive. The factory work gives them increased power, which makes it in fact desirable from a feminist point of view, as it is putting these women in a more equal position, even though the conditions are bad. This suggests that not shopping in H&M would actually be a bad thing, as it would in fact cause harm by taking these jobs away, along with that freedom and power.

Yet, that position is also not satisfying. Fraser's argument makes Pettinger consider that, in fact, Kabeer's feminist position that such work is empowering is a mistake (a 'failure of feminism') as it plays into the hands of exploitative capitalism by making unsafe, badly paid work something positive, and even to be celebrated ('a new romance of female advancement'). That is a mistake because it makes what is effectively a different type of oppression seem like liberation.

In the final paragraph, Pettinger considers both of these arguments side by side. These two positions – Kabeer's and Fraser's – are contradictory and seem to both be valid. The terrible conditions in the factories that cause injury and death 'is damage'. Not shopping in H&M because of this would also, however, cause harm to these women as they would lose their jobs and the associated autonomy and agency ('This is damage too').

What, then, is Pettinger's final argument? It can best be summed up in the quote: 'Strong conclusions are impossible when there's only a choice between forms of damage (Pettinger and Lyon, 2012) In other words, Pettinger's final argument is that there is no possible correct choice here – either way causes damage. However, this is still an argument that has been explained very clearly, and her 'playing out [of] the tensions' has illuminated the paradox very well. Reading this enables us to understand clearly the costs and benefits of either choice – to shop, or not to shop – and thus informs an ability to 'choose well'.

Reflecting on reading

From this account, we can see how much argument was condensed into this piece that, on the surface, seemed fairly simple. We can also see how much more productive it is to read actively and not passively.

The final step here would be to reflect on the reading that we have just undertaken and to continue the conversation with the text. We cannot do this in the way that we would do with a text we encountered as part of our studies, as there is no assignment, module or course to contextualise it within. You could, however, pause to consider the answer you gave to the ethical dilemma of whether or not you should buy goods produced in poor conditions a few pages ago, and see whether anything has changed through your reading of Pettinger.

More useful, however, would be to reflect upon your experience of reading this chapter within the context of your reading of this book.

TASK 4.6

Think about the following question – what has reading this chapter taught you about how arguments work? Take notes on how reading this has changed your understanding of arguments, and try to identify three concrete things that you will do the next time you encounter an academic text to make sure you are reading actively and attentively.

1.
2.
3.

Summary

- Active reading is close, attentive reading that engages with and questions the text being considered. This is the best way to ensure that you are reading for argument as well as content and is vital to academic success.
- You should always be reading with a clear purpose in mind, and be reading strategically. Try and work out what you want to get out of reading a text, and set yourself questions that you want to answer before you start reading.
- Ask questions as you read, and make sure you are always trying to understand what arguments are being made in the text, as well as what information is being conveyed.
- Take notes as you read. Note-taking is one of the most important ways to read actively, and to make sure that you understand the argument a text is making.
- Reflect on the text after you read to see what you have understood from it, and whether it has answered your questions.
- Reading will tell you not only about the content of those texts, but about what makes a good argument in your subject. How are they structured? What sort of evidence do they use? What sort of claims do they make based on their conclusions?
- Everything here applies equally to listening as it does to reading. As a student, you should be looking to 'read' every lecture, discussion, or debate in the same way you would a written text.

FURTHER READING

Chapter 2: Spelling out Arguments and Assumptions from Chatfield, Tom, *Critical Thinking* (Sage, London, 2018).

Prose, Francine, *Reading Like a Writer* (Aurum Press, London, 2006).

Chapter 2: Reading Arguments from Ramage, John D., Bean, John C. and Johnson, June, *Writing Arguments: A Rhetoric with Readings*, 5th edn (Allyn & Bacon, USA, 2001).

5

Counterarguments

A key part of engaging with and producing arguments is the ability to test and validate propositions, whether they are your own or others. Thinking about possible challenges and alternatives is a key aspect of this. The process of generating counterarguments is a vital part of the whole academic process – we do it when engaging with the arguments of others, when producing our own arguments and when we consider any position, assertion or assumption.

Counterarguments are not necessarily adversarial in nature. Just as the term 'arguments' in an academic context does not only refer to conflicts or debates, but also specific positions taken in response to particular questions, so 'counterarguments' in an academic context are not just about proving someone else wrong or 'winning' – although that can often be the purpose. Depending on the situation, counterarguments can be about improving upon an existing position, showing that another way of thinking or proceeding is possible, correcting misunderstandings and gaps in evidence, or even just testing an argument or assertion to see if it is valid or has any flaws or issues.

For example:

Argument – Racial equality has still not been achieved in the UK.

↓

Counterargument – The 2010 Equality Act guarantees equality in law and there is much evidence to suggest equality has been achieved, including the most diverse ever parliament in 2019 and much more numerous cultural and supporting figures of colour.

↓

Response to counter – The fact that it was necessary to pass legislation protecting equality as late as 2010 shows precisely that equality has not been achieved, as if it had, there would be no need for the law to be updated in the first place. Likewise, the make-up of parliament still does not reflect the demography of the country as a whole, and the existence of sporting and cultural figures does demonstrate some progress, but not equality.

↓

Adapted argument – *Despite some improvements,* racial equality has still not been achieved in the UK.[1]

> **Synthesis** – Bringing together different elements to produce a new argument that is both reflective of and different from the constituent parts.

We will look in more detail at how counterarguments can be integrated into and used to help structure arguments in Chapter 6. In this section, the focus will instead be on how we generate counterarguments, and use them both to test and evaluate the arguments of others and to produce our own arguments.

In order to do this, this chapter is going to use a framing question as a context. That question is: are you a feminist?

This may be a question that you have given a lot of thought, or it may be a question that you have never considered. Likewise, it may be a term that you are familiar with or one that you are unsure about. However, it is likely that you have some understanding of this label and that you have an immediate response to the question. Therefore, whatever your level of familiarity with the idea, take notes on the question for a few minutes, and think about what is affecting your answer.

QUESTION: Are you a feminist?

TASK 5.1

Take notes on or think about this question and the reasons that support your answer. Think about any issues or experiences that might be shaping your view and how you feel about the question. Does the term provoke a negative or positive response? Why?

ANSWER: Are you a feminist? Yes or no?

It is likely that you had an instant response to this question, whether positive or negative. It is also likely that in thinking about your response you had to answer a lot of other questions. For example: what does it mean to be a feminist? What do I already know about this term and how do I feel about it? Is being a feminist a good thing or a bad thing? Why might some people argue that it is or isn't? How do I feel about those people? What debates or issues in wider society does this question relate to? Can a man be a feminist? And so on.

As we saw in Chapter 1, our response to any question will necessarily involve asking and answering a whole range of associated questions, and at times we may find that our response is actually to one of these secondary questions rather than the original.

You might also have considered some of the questions from Chapter 4 here, or a version of them – namely, why am I being asked this question here, in this context? What is the expected response? What does this have to do with building arguments at university or generating counterarguments?

One answer to that question is that, as we have already seen, we can only get so far by discussing arguments in the abstract. In order to usefully engage in any exploration of how arguments and counterarguments work, we need to engage in real issues and questions. This particular topic – feminism – is intended to be accessible, both to a wide range of individual backgrounds, and also to be relevant to a wide range of disciplinary interests. That is, this is a question that a large number of people should be able to engage in without any specialist prior knowledge, and it is a question that should then usefully relate to their lives and future studies. Whether that supposition is correct, I will leave you to decide.

In other words, we are considering the question 'are you a feminist' as a context in which to explore how we can generate and respond to counterarguments.

Having clarified that, let us consider the overall question and the subordinate questions we have generated to help us think about it.

What does it mean to be a feminist?

There are multiple ways to answer this question, but at its most simple, feminism is the belief in gender equality. What exactly that means depends on how you approach the question. Various versions of feminism, usually referred to as 'waves', have emerged over the last century or so, concerned with, variously, women's political and legal rights; the social roles of men and women; reproductive rights, sexuality and gender-based violence; exploring the ways in which gender roles are not 'natural' but are rather constructed socially; and seeking to either change existing social systems, or break them down all together.

As we saw in Chapter 4, an ideology (according to one definition) is a set of political ideas with a clear medium- to long-term goal. In the case of feminism, that goal is gender equality. Different waves involved different specific goals, different ways of achieving those goals, and different theoretical focuses, but in general the founding principle is the same – equality.

If you believe that men and women are equal and should be treated accordingly, therefore, you are a feminist.

What does all of this have to do with building arguments at university?

In most Western contexts (the term 'Western' is a geographically inaccurate one usually used to cover Europe, America, Australia, and assorted ex-colonial outposts) you would expect the average university student to consider themselves a feminist. After all, this is a fairly standard liberal opinion, and at its most basic, seems difficult to object to.

Whenever you were directly asked this question in an academic context, therefore, it is likely that the expected answer would be 'yes'. But it is worth noting that there are plenty of places in academia where this question would *not* be asked, and the dominant position would not be feminist – although this would be likely to be an implicit rather than an explicit position that suggested that feminism was not relevant to the matter at hand rather than that it was actually wrong.

Consider, for example, how history tends to be dominated by male-focused narratives and authors (see Kahn and Onion, 2016), and how women's history is an addition, a subset, in the same way as black history is in the UK or US. That is, we talk about 'women's history' but we do not talk about 'men's history', as we do not feel the need to. Men are the norm, and women are therefore different from and subordinate to that norm.

The sciences are equally affected, with medical education texts, for example, tending to use the male form in case studies and anatomical drawings, and only including representations of women when discussing reproductive anatomy (see Parker et al., 2017). This gender bias is reflected in the practice of healthcare professionals and in safety standards in industry, where products are designed based on the male anatomy as a standard. This shows how many things that are often considered 'facts' (e.g. 'this car is safe as it has been scientifically tested') are often actually a result of a series of choices and assumptions, and that scientific knowledge (e.g. medical science) is also affected by social and ideological factors, and cannot be considered separate from them.

The university as a whole, and the subjects of study that make up that institution, at least in the developed world, exist within societies that remain predominantly male and white, in worldview if not in actual physical make-up. The knowledge that is produced by the university and the disciplines within it is shaped by that social reality, and feminist thought in all subjects aims to tackle not only the wider issues of gender equality, but also the ways in which this is built into knowledge and knowledge production. This is why almost every area of academic thought has a feminist strand.

This is another reason why you are being asked this question here. Plenty of things that don't look like feminist issues *are* feminist issues, and it is as worth thinking about when this question is *not* asked as when it is. The same is true of the increasingly common discussions around **decolonising the curriculum**, where many issues that seem to be unrelated to race and colonial history are in fact deeply enmeshed with them.

Decolonising the curriculum is a concept concerned with paying attention to how knowledge is and has been produced, and attempting to approach that knowledge through new perspectives and create new connections to allow different voices and positions to emerge. This is not about deleting existing curricula or histories, but rather opening them up to new ways of knowing and enriching our understanding (see, for example, Arshad, 2021).

What do I already know about feminism and how do I feel about it?

The answer to this question will depend upon a lot of factors, including which country you live in, your social, economic, class and gender background, and perhaps especially your age. Whatever you think about this topic is likely to be different to your grandparents, and perhaps even your parents, even as what they think is likely to be very different from the generations preceding them.

Globally, however, there are certain common features that are likely to be consistent. A 'backlash' against feminism has occurred in various forms around the world over the last 40–50 years, arguing that feminism has 'gone too far', is harmful, or undermines valuable traditional forms of social structure and knowledge. At the same time, events such as the

'Me Too' movement, or the struggle for female education in the developing world (think of the global fame of Malala Yousafzai, for example), consistently occur to demonstrate the continuing need to address gender inequality.

It is likely, therefore, that you will have been exposed to both positive and negative representations of feminism in your everyday life. It is worth noting, however, that the assumption that feminism is widely accepted in a university context is itself questionable. Various pieces of research over the last few decades (see, for example, Houvaros and Scott, 2008, or the popularity of figures such as Jordan Peterson) has shown that students – whether male or female – are much more reluctant than you might expect to identify as feminist, and that they do not always see the label, or the idea, as something positive.

As an example, at the University of Essex in 2017, the (student) Feminist Society held a bake sale to highlight the gender pay gap – i.e. the fact that women in the UK at that point were, on average, paid 18 per cent less than men. As part of this, they sold cupcakes for £1 to male students, and 82p to female and non-binary students, billing this as a way to redress the balance. In reality, it was about raising awareness of the issue in an eye-catching way. Another student reported this as a 'hate crime' and claimed it was discriminatory (Gray, 2017).

Whatever your position on all of this, however, it is worth noting that none of your responses to any of these questions is 'natural'. You think these things for a reason, and you have absorbed lots of arguments about feminism, or at the very least, what is 'true' about male and female roles in society, and the 'nature' of men and women, whether that be in terms of the way they think, the way they behave, the way they experience emotion, the way they conduct relationships, or the way that they have or desire sex. These arguments will have fundamentally shaped your assumptions about how the world works, and will therefore shape your response to any questions about the issue, whether you realise it or not.

This is as true of your response to any question as it is to a question about feminism, and it is important to remember these assumptions in your own thought, as well as looking for them in the thought of others.

For the purposes of this chapter, we are going to assume that the answer is 'yes' – you are a feminist, and you think feminism is necessary and correct. That is the position that we are going to take in response to the questions and texts that we will encounter here. Bear that in mind as we continue, as it will affect the way that we interact with different positions and propositions.

Logical fallacies

Before we look at an example text to further explore this question of feminism, it is first necessary to consider a few of the ways in which we can generate counterarguments.

Firstly, we can think about counterarguments as a way of testing an argument, just in terms of how well it stands up under its own terms. That is, in order to evaluate how strong an argument is, we can ask ourselves – are there any problems with the argument itself and the way that it has been put together?

> If all the premises in an argument lead logically to the conclusion, then that argument is **valid**. If the premises of the argument are also true, and the conclusion is therefore also true, then the argument is described as being **sound**. See Chapter 2 for more details.

The easiest way to do this, in the first instance, is to consider whether there are any mistakes or flaws in the argument. We looked in Chapter 2 at how to think about whether an argument is **valid** or **sound** in general terms, but here let us think more specifically about some general types of poor reasoning that can lead to wrong conclusions. Some of these are called *logical fallacies*, and being aware of them can help us to spot a problem with an argument that otherwise seems convincing.

Here are some common examples:

Factual error – This may seem obvious, but it is important to consider. Errors can range from the outright mistake (or falsehood) to the misinterpretation of data or the misapplication of a theory, but it is always worth noting that someone might simply have got something wrong, and spotting that mistake can make you realise that the whole argument is unsound. This is as important to check in your own work as it is in the work of others.

Ad hominem – This is a type of logical fallacy, very common in politics, and literally translates as 'against the man'. In practice, this means attacking the person making the argument in order to suggest that what they are saying is wrong. 'Don't listen to him, he's an idiot' is the simplest form of this attack.

Straw man – Constructing an exaggerated and deliberately weak position in order to knock it down and make the opposite seem more convincing. This is named after the straw-filled dummies that were used to practise sword-fighting – in other words, this is something that looks like an enemy but isn't really at all and 'beating' it proves nothing.

Tu quoque – This translates as 'you too', and refers to the accusation of hypocrisy, or someone doing the very thing they are criticising in someone else. In other words, it is saying – you do exactly the same thing, or your argument has exactly the same flaw. The problem is, of course, that while the accusation that someone else is just as bad might be true, it doesn't tell us anything about the quality of the original argument. In fact, it sometimes suggests that the person crying '*tu quoque*' knows they are wrong and is sticking with their argument anyway ...

Begging the question – Tautology and begging the question are two ways of thinking about the same thing – they are both arguments that have conclusions based on their own premises and so are unfalsifiable. Some are very obvious ('cigarettes are bad for you because smoking kills'), but others are much less so. This is also sometimes referred to as *circular reasoning*, as the premise assumes the conclusion, or *tautology*, where the 'conclusion' is just a restatement of the premise(s) in another form.

Equivocation – This fallacy occurs when the same word is used to mean different things, whether accidentally, deliberately, or because of an ambiguity of definition. A famous example of this is the US President George W. Bush's declaration in 2006 that 'The US does not torture'. This statement was only true to a given definition of the word 'torture' that excluded many acts considered torture under international law, such as waterboarding, which the US *were* doing.

False dichotomy/false dilemma – Suggesting that something is true by creating a false sense that there are only a limited number of options available. Again, this is common in political discourse ('if Russia does not retreat, then war is inevitable'), but is a common fallacy that can be much more difficult to identify. This can be related to *over-generalisation* – for example, if the outcome of a single experiment is used to argue that a hypothesis is or isn't proven, or if an individual event is used as a model for all possible situations.

Appeal to ... – There are many outside factors that can be appealed to in order to guarantee or justify an argument but the most common are ignorance and authority: An *appeal to ignorance* is where we assume something is or isn't true simply because the opposite hasn't been proven yet, e.g. 'there can't be alien life in the universe because otherwise we would have found it if there was'. Of course, just because we haven't yet proved something to be true does not mean that it is false.

An *appeal to authority* argues that something is true simply because of the source it comes from (e.g. 'I know that this article is good because my lecturer recommended it'). Note that this is not necessarily a fallacy – authorities are often correct and should be looked to for information on their subjects of expertise. However, being an expert, or other form of 'authority' does not mean that someone should automatically be believed, and we should also ensure that people are authorities in the field we are actually considering – a very important reason to be cautious of celebrity endorsements, for example.

There are lots of other things we can appeal to – *popularity* (10 million listeners can't be wrong!), *tradition* (we've always done it like this!), *nature* (men are naturally more assertive so more likely to succeed in business) and even *sympathy* (the death penalty is justified because victims of horrible crimes deserve justice). Many can be convincing, some can even be correct, but we should still be careful any time that some outside factor is appealed to – this can distract us from the validity of the argument itself.

TASK 5.2

Take a look at the following examples, and see if you can identify the problem in each case:

1. Smith (2018) criticises the study for only using a small number of participants from one social group. However, this is a standard feature of such experiments and so the criticism can be ignored.
2. Life expectancy in Africa has risen astonishingly as that country has entered the global economic system.
3. Jones is a Marxist, and so their reading of Ayn Rand is unlikely to be useful.
4. Free trade will be good for the national economy as unrestricted commercial relations will give all sections of the country the benefits that result when there is an unimpeded flow of goods between states (adapted from Engel, 1994).
5. Those arguing that gender is socially constructed deny that there is any biological basis for sex differences. Such anti-scientific perspectives should not be taken seriously.

ANSWERS

1. This is an example of a *tu quoque* fallacy. The existence of other poorly designed experiments does not justify poor sample selection in this study,
2. This is a *factual error* – Africa is not a country. Boris Johnson said this while Foreign Secretary of the UK in 2016.
3. This is an example of *ad hominem*. Jones being a Marxist may make them likely to disagree with Ayn Rand's position, but that does not mean that their argument will be incorrect or of no use.
4. This is an example of *circular reasoning* – the conclusion is just a restatement of the reason given for it.

5. This is a mixture of a *straw man* argument, an *ad hominem* attack and *equivocation*. Arguing that gender is socially constructed is not related to biological sex distinctions and so is a *straw man* argument; this inaccurate suggestion that gender and sex are synonymous terms is *equivocation*, and suggesting that such viewpoints are 'anti-scientific' is an *ad hominem* attack.

Generating counterarguments

Checking for any mistakes or faulty logic is just one way of testing arguments. There are lots of other things that we can do to generate counterarguments and see what other positions are possible or what improvements can be made to a line of reasoning. These strategies apply just as much when you are considering your own arguments as they do when you are examining the arguments of others.

Here are a few for you to think about.

Think like a believer

The first step is to consider how you, as a reader, are approaching the text. Perhaps counterintuitively, in order to think about challenges and alternative points of view, you first need to try and really understand the argument that someone is making. To do this, you can 'think like a believer' (Ramage et al. 2001, p26), and try what Carl Rogers (2017) called 'listen[ing] with understanding' (see Chapter 2 for a more detailed discussion of this). In other words, you try to really inhabit the position being taken, both in terms of its claims and all of the values that underpin it, 'to really achieve ... [their] frame of reference' (Rogers, 2017). This will allow you both to properly evaluate the argument and to really understand how it has been put together.

There are various ways you can help yourself think or read like a believer, and one of the key ways of doing this is to attempt Rogers' thought experiment (see pp29–30) – namely to summarise the point of view so well that the person arguing it would be satisfied with your version. Doing this will force you to properly and fully understand what you are dealing with and enable you to consider it critically.

Think like a doubter

The flipside to thinking like a believer is, of course, thinking like a doubter and coming up with questions, challenges and reasons to be sceptical.

There are a number of questions you can ask yourself here, including the following.

What paths were not taken?

Answering any question and building any argument involves taking a certain approach. How the question is framed, why the question is being asked, what evidence is used, any hypothesis that will be tested – deciding all of these requires not only choosing a particular path, but also means that other paths cannot be taken. In economics, there is the idea of **opportunity cost**, or the benefits that you would have gained from the option that you didn't take, and a similar effect is relevant here. Considering what other paths could have been taken in an argument is one key way that you can generate counterarguments, whether to challenge a given position or strengthen and complexify it.

> **Opportunity cost** – The benefits that you would have gained from the option or options that you didn't take and that have therefore been lost.

What evidence is being used here and why?

All arguments are backed up by some form of evidence, whether that be experimental data, historical fact or secondary quotation. When considering this evidence, as well as checking for any errors or logical missteps, it is also important to consider why *that* evidence is being used and not something else. In any given situation, the person making the argument will have to have made decisions about the evidence they are going to use to back up their position, and those decisions are not neutral. Thinking about the evidence used will help you to understand those decisions, and what has shaped the approach taken, even if that is not explicit.

What's missing?

Any choice of what evidence to include necessarily also involves a choice of what to exclude. Sometimes this is conscious, but either way, it is interesting to consider what *isn't* being said in any given argument and why. Considering both what has and hasn't been included allows you to think about the impact of alternative points and whether any important omissions have been made.

What method has been used and why?

Often, the evidence used or omitted will have been determined by the method used. In the sciences and social sciences, this will often be very explicitly discussed, and critiquing the methodology used in any piece of research (and the arguments resulting from it) is a standard part of academic practice, and will be explicitly taught in both undergraduate and postgraduate study. However, in any discipline, a 'method' of sorts has been chosen and applied, even if that decision is more implicit. For example, when examining a novel, a literary scholar could use a biographical, new historicist or post-structrualist approach (among others), while a historian could, for example, approach any given event from a socio-cultural, economic or ethnohistoriographic perspective. Each will have an effect on the outcome, and considering the results that alternative approaches might have produced will help you to consider possible counterarguments or ways of thinking.

How would thinking about this from another perspective change things?

Underpinning all choices about evidence and method are theoretical frameworks and approaches that decide how questions can meaningfully be answered. Some of these are disciplinary – think, for example, of the differences between how a biological scientist, a philosopher and an architect might approach the question of what constitutes beauty – while others are ideological, or simply more pragmatic. Choosing one or the other is not a value judgement – psychology is not 'better' than sociology, and new historicism is not 'better' than deconstruction. Different forms of knowledge, different forms of 'fact' or 'data', are the base units from which we construct arguments, and we create these different forms by choosing these different approaches.

As we explored at the beginning of this chapter, in order to answer a big question we often focus down onto smaller, more manageable questions and explore those, and there is often some distance between the evidence used or facts deduced and the argument made based on those facts. Thinking about how the counterarguments that different approaches might suggest, and moving between the different levels of abstraction (i.e. between a detailed consideration of individual pieces of evidence and a broader interrogation of context and approach) are both vital parts of testing any argument and improving it.

Here, it is also necessary to think about the purpose of an argument or the underlying assumptions. Does it relate to an existing debate or way of thinking and is it trying to achieve a particular goal?

The further you go through your studies, the more you will be aware of different approaches within your own discipline and how they help determine arguments. But it is also worth being open to ideas and concepts from other disciplines and forms of study as these can help to inspire different ways of thinking and stop you from seeing the processes and frameworks of your area as being the only way of doing things.

Can you counter the counter?

Having thought of possible objections, questions and alternative ways of approaching things, you can now ask yourself – how would you respond to these counterarguments? This may require further research or thought, but considering counterarguments can be a productive way to both improve your own understanding of, and engagement with, another's argument, or to more fully understand your own. See the example at the start of this chapter (pp87–88) for an illustration of this.

TASK 5.3

Take a look at the following examples and see what counterarguments you can come up with. They are all quite brief so it may not be possible to do all of the steps above, but try and come up with as many challenges, problems and potential questions as you can.

1. People who live in Mediterranean countries have fewer heart attacks than those who live in Northern Europe. The diet in those countries includes a lot of olive oil – therefore, eating olive oil makes you less likely to have a heart attack.
2. During the Brexit campaign in 2016, when people in the UK were asked to vote on whether or not they wanted to remain part of the European Union (EU), the following claim was made by the Vote Leave campaign: 'Turkey (population 76 million) is joining the EU'. Suggestions were made that between 5 and 15 million Turks would come to the UK as a result in the first five years. What do you think of the claim that millions of migrants would be headed to the UK if Britain stayed in the EU?
3. Paying unemployment benefit makes being out of work attractive. Therefore, lowering the level of benefit – i.e. the amount of money people receive from the state when they don't have a job – will incentivise them to find work and lower the unemployment rate.

ANSWERS

1. This sort of claim about diet and nutrition is very common and is frequently reported in both advertising campaigns and the media. It is an example of *correlation not equalling causation* (see Chapter 2), and more broadly we can see that it relies on taking two separate facts – incidence of heart attacks and presence of olive oil in the diet – and not considering any other factors that might be relevant. Also missing here is a consideration of anything that might be *causing* heart disease in Northern Europe or any other contextual detail.
2. A quick check tells us that currently, in 2022, Turkey has still not joined the EU. Even ignoring this fundamental problem with the argument presented by Vote Leave, we can see that the number of suggested migrants as a proportion of the population of Turkey is very high and thus unlikely to be realistic. A consideration of why this argument was made is also important here. The idea was to promote fear of large numbers of immigrants, and the fact that it was possible to suggest this simply by stating the population of the suggested country shows how prevalent anti-immigration sentiment was at the time.
3. There are a number of things to think about here. Firstly, the claim that lowering unemployment benefits leads to lower unemployment figures is one that could be checked against available data. The claim that receiving unemployment benefit makes being out of work attractive is also questionable and evidence for this would need to be examined. The rest of the argument fundamentally rests on this claim, and if it were found not to be true, then the rest would also not be true. Other potential challenges include the fact that if there are not enough jobs for people then any incentive to work will have no effect, and the impact that lower benefits might have on the ability of individuals to find a job (e.g. being able to travel to interview, having the necessary clothing and equipment). This argument is common, and rests on a particular theoretical framework – that is, a conception of poverty and worklessness as the responsibility of the individual. Starting instead from a point of view that saw employment and welfare benefits as the responsibility of the collective (i.e. society or the state) would be likely to lead to a very different position.

Putting it into practice

Having thought about all of the possible strategies for generating counterarguments, let's put this all into practice by returning to our topic – feminism – and in particular, analysing a short extract from a text that represents a challenge to the pro-feminist position that we have decided to take in this chapter.

The short text that we are about to look at is taken from the Prologue to Neil Lyndon's 1992 book *No More Sex War: the Failures of Feminism*. Lyndon is a journalist rather than an academic, and while in many ways the book is obviously outdated, it remains representative of certain key arguments from the backlash against feminism, and the fact that Lyndon republished it in 2014 suggests that he considered its arguments to remain relevant.

You do not need to know anything about feminism, in the UK or elsewhere, beyond what has already been discussed here to critically approach this extract. If you would like to, please revisit *the trigger questions from Chapter 4* before reading.

In the prologue to his book, Lyndon (1992) argues feminism is a failure that is based on fundamentally wrong ideas and has had an actively negative effect on society. He says that feminism has 'declared a war of eternal opposition between men and women' (p2), and attributes the 'conspicuous failure' of his generation (i.e. Baby Boomers, those born in the decades after the Second World War) to deliver 'radical change in the institutions of state, in the rights and freedoms of individuals', to 'the influence of feminism and to the perverted account of personal relations and of social composition which feminism has fostered' (p2).

To support these claims, he gives the following 'short list of facts'. While reading these, use the approaches above to see what you think of Lyndon's (1992) argument and see what objections you can come up with.

Think:

- What evidence is being used? What is missing?
- Does the evidence support the conclusions drawn? Are alternative conclusions possible?
- Can you spot any logical fallacies, errors or examples of faulty reasoning? For the purposes of this exercise, you can consider all factual assertions here to be correct.

Here is Lyndon's (1992) list. I have numbered the points to make things easier:

1. Until the autumn of 1991, when the Children Act became law, the fathers of at least one in every four children born in Great Britain had no rights with regard to those children.

Approximately 175,000 children are born every year to unmarried women. Those women have had all the rights of parenthood for those children. The men have had no legal rights of paternity ...

The Children Act is intended to afford to unmarried fathers the right to acquire parental responsibility on the same terms as married fathers; but we cannot, at present, guess how its vaguely expressed terms will work in practice

2. A man who makes an application to the divorce court for joint custody of the children of a broken marriage has a one-in-five chance of success. A man who makes an application for sole custody of his children has a one-in-ten chance of success.

About 175,000 divorces are granted in Great Britain every year. In more than 100,000 of those divorces, the couples have children under the age of 16. The routine practice of the divorce courts of Great Britain is to strip men of their property and income and, simultaneously, deny them equal rights of access to and care for their children.

3. About 200,000 abortions are legally effected in Great Britain every year. We do not know how many of the fathers of these foetuses might have wished to see their children born: nobody has ever tried to count them. They are not accorded a glimmer of public attention nor an atom of legal rights. Subject only to the consent of doctors, the pregnant woman is given the absolute right to choose to abort the foetus, regardless of the father's wishes or the state of their relationship when she conceived. The inseminating man has no right, in law or convention, to express or to record an opinion on the abortion, even if the woman has previously openly and unambiguously expressed their desire to bear their child.

4. About 700,000 babies are born in the UK every year. The 700,000 men who are their fathers have no right in law to time off work when those babies are born.

5. A man may not be classed as a dependant for social security benefits in Great Britain.

6. Widowers who are left with the care of children are not entitled to the state benefits which a widow would receive.

7. Though it will soon be changed, the law in Great Britain still allows women to retire and receive a state pension at the age of sixty while requiring men to work until they are sixty-five. The coming change in the law has been imposed upon Britain to bring the country into line with its European partners. The protests of men on this incontestable point of inequality have been ignored in Britain for twenty-five years. (pp3–5)

Lyndon (1992) says that, 'two connected consequences flow from this list. First, it gives an unusual perspective on our times and the societies in which we live ... Second ... it must mean, in each of its particular parts and in sum that the cardinal tenets of feminism add up to a totem of bunkum' (pp5–6).

For him, the chief of these tenets, 'the corner-piece of that mosaic of belief, assertion and argument [i.e. feminism] has been the claim that all post-nomadic societies have been patriarchal ... that those societies have been organised by men for the benefit of men and to the disadvantage of women' (p7). He argues that, 'If any disadvantages apply to all men, if any individual man is denied a right by reason of his gender which is afforded to every single woman, then it must follow that ours is not a society which is exclusively devised to advance and protect advantages for men over women. It is not a patriarchy' (p9), and claims that the above list of facts therefore provides proof of the fact that UK society is not a patriarchy. Feminism is therefore wrong, and is attempting to address a problem that does not exist.

Before we try and see what counterarguments there are to Lyndon's position, let us first try to *think like a believer*.

TASK 5.4

Write a short summary (i.e. less than 100 words) of Lyndon's position that you think he would be satisfied with.

Having done this, let's return to the question I asked you at the very beginning of this chapter. Are you a feminist? Has reading Lyndon's (1992) argument changed your mind?

QUESTION: Are you a feminist? Yes or no? Why/why not?

Now, having tried to fully inhabit Lyndon's position, let's *think like a doubter* and see what possible counterarguments there are to his arguments.

We have a list of seven 'facts' to consider, which are used as evidence to support an overall argument that feminism is wrong because it is based on the erroneous assumption that society is patriarchal.

Before we interrogate the overall claim, let us take each of the 'facts' in turn and as an overall list. What problems and challenges can you come up with in each case? Can you spot any flaws? Why has this evidence been chosen and what is missing?

TASK 5.5

Look back at Lyndon's list of 'facts' and take notes on the above questions.

Before considering the overall list, let's look at each point in turn:

Point 1 – This is clearly unfair. However, this is a recognised injustice that has already been addressed (through the Children Act), as Lyndon himself recognises.

What is missing here is a sense of how many of these unmarried fathers are being denied any rights – that is, how many of the 175,000 children have fathers that are not being allowed to do something they want to – and a consideration of historical context. Children being born outside of marriage was only normalised from the 1960s onwards in the UK, and men not having 'rights' to such children could arguably have been seen as a benefit, in that they were not required to take responsibility for something unwanted, rather than that they were being denied access to something desired. Existing legislation and parental rights were therefore reflecting this norm, rather than designed to disadvantage men. Laws are made to address specific issues and behaviours and are born out of specific social contexts – they are not necessarily expressions of moral positions.

The flipside not mentioned by Lyndon (1992) here, related to this, is the fact that as these fathers were not recognised by law, the mothers also had no ability to impel the fathers to assist in raising their children. In this sense, their legal 'right' to the child is also a legal *obligation* to care and raise that child, which was not incumbent upon men.

Point 2 – At first glance, this also seems like an injustice, and anyone being unfairly denied access to their children is wrong. However, there are several issues here. Firstly, note the emotive language ('strip men of their property ... deny them equal rights'), which seems designed to provoke an emotional response rather than present an objective account. Secondly, we don't know enough about any of these claims to say that this is 'routine practice' and not a justified response based on the cases at hand; thirdly, joint or sole custody are not the same as 'equal rights of access and care' and there are many forms of arrangement that would constitute equal access but not joint custody. To an extent, Lyndon (1992) is guilty of creating a *false dichotomy* here.

Lastly, the property of a marriage and the income of those within it are seen in law as shared, not the sole property of the man. The routine practice that Lydon describes here is to split that property and income between the parties, so men are not being 'stripped' of their assets or income, but rather being made to divide them fairly. Lyndon is unintentionally showing the patriarchal basis of his thinking here, both in assuming that it will be the men that have the income and property, and that taking any of it away constitutes some sort of theft.

Also missing is a sense of what proportion of men apply for joint or sole custody, and the proportion of men who do not want custody or do not fulfil their legal obligations (e.g. paying child support) following separation. Without this, we can't contextualise the figures or truly draw a conclusion as to what they mean.

Point 3 – As with point 2, Lyndon (1992) uses very emotive language here, which is problematic given the fact that abortion is a hard and recently won right and a very emotionally freighted topic. Women being denied the right to control their own bodies or reproductive capabilities is a major aspect of most of human history and continues to be a very contested battleground. That context is ignored here. Women being abandoned by men to raise unwanted babies is also a very real problem – as is women being forced to have abortions, legal or otherwise. The legalisation of abortion in the UK in 1968 is widely regarded as one of the 'radical change[s] in the ... rights and freedoms of individuals' that Lyndon (1992) wants his generation to have made, so it is important that he does not mention any of this context.

Last but not least, and as with point 2, a fundamental question not answered here is – how much of a problem is this? If there were large numbers of men feeling that their unborn children were being stolen from them, it is hard to believe this would not have been considered, especially when anti-abortion stances are so politically popular. Lyndon (1992) himself says that no one has tried to count this, and implies that it's because no one cares for these men. It could equally be because there is nothing, or very little, to count.

Point 4 – Lyndon is correct – this is/was a scandal. Missing, as with the previous two points, is a sense of how many men actually *want* such time off, but overall he is right to highlight this issue.

But can we counter this counter? Yes. Men being denied the right to time off after their children are born means that the burden of childcare falls almost completely on women and that they have no choice in the matter. Something being bad for men doesn't always automatically make it good for women, or vice versa – a point that Lyndon recognises but does not always seem to follow to its logical conclusion. In this case, both men and women are disadvantaged by the law rather than a disadvantage for men being an advantage for women. The perceived 'battle' between men and women that Lyndon (1992) deplores seems to have infected his own thinking, and what he sees as an example of men being disadvantaged is in fact a feminist issue that affects both sexes.

Point 5 – again, this is wrong. But it is also a legacy of patriarchal society – this is because men were (and still are in many cases) presumed to be the 'breadwinner' (that is, the one responsible for 'providing for' the family) and always in charge of money. Women in the UK were traditionally denied access to money and *forced* to be dependent, whether they want to be or not. Thus the law reflects historical and still existent social norms, where men cannot be considered dependents because it is assumed that there is no way that they *could* be dependent on a woman.

Patriarchy is not always good for all men or what all men want. Something that disadvantages some men can be a result of something that also disadvantages women. Again, the system here disadvantages both men and women, and so this is also a feminist issue rather than an illustration of a failure of feminism.

Point 6 – Again, this is a legacy of the patriarchal assumption that a widower will be the breadwinner and so will have an income after their partner's death. Widows received benefits as it is assumed that they would not have such an income and would therefore otherwise be left with no money. There are also assumptions about childcare and who carries it out here.

Missing again, therefore, is a consideration of the wider context of financial and economic rights and benefit allowances. We cannot conclude whether the system as a whole advantages women over men, or vice versa, based solely on this point.

Point 7 – This is a valid point. However, it is also 'soon [to] be changed', so, as with point 1, it is a recognised injustice. It is therefore arguable whether the 'protests of men on this incontestable point of inequality have been ignored in Britain for twenty-five years', or whether the protests have in fact been listened to and led to a change in the law.

The whole list

Having considered the points individually, let us now look at the list as a whole. The evidence used has clearly been chosen in an attempt to show that there are a number of ways in which men are disadvantaged compared to women. Some of them are valid and do suggest that men are disadvantaged in particular ways.

However, there is too much missing here to draw strong conclusions. We do not have a comparable list of inequalities or injustices affecting women, and in most of the points Lyndon (1992) lists we are also missing key contextual details that make it difficult to judge whether his interpretation is correct.

Also missing is a sense of *why* any of the laws or situations quoted are the way they are, or the motivations behind them. Lyndon (1992) frames this list as if the points included are indicative of conscious attempts to disadvantage men, yet does so without any of the necessary historical context.

Let us consider the claims that Lyndon (1992) makes based on this list – that society is not patriarchal and that feminism is thus rendered null and void.

Firstly, is it true that, 'if any disadvantages apply to all men, if any individual man is denied a right by reason of his gender which is afforded to every single woman, then it must follow that ours is not a society which is exclusively devised to advance and protect advantages for men over women. It is not a patriarchy' (p9)? No, it is not. This is another example of a *false dichotomy*, and to a certain extent of *equivocation* or even of a *straw man* argument.

A very basic definition of patriarchy, taken again from simply searching on Google, is that it is, 'a system of society or government in which men hold the power and women are largely excluded from it'. The word 'largely' is key here – and note how this contrasts with the 'exclusively' used in Lyndon's (1992) definition. Lyndon's points, at best, show that there are some examples in which women are treated better than men by society, but this is not enough to discount all of the other ways in which society does the opposite.

Lyndon's (1992) overstatement of what patriarchy means is *equivocation* and he uses that equivocation to set up a *straw man* argument. If feminism did really argue that patriarchy was about a society that *exclusively* gave power to men, then it becomes a very easy argument to knock over, as all one needs to do is find one example of where women seem to hold the advantage – the one black swan. Feminism does not argue this, and so the whole premise of the prologue is a *false dichotomy*.

To return to the very beginning, we also need to consider Lyndon's (1992) overall definition of feminism as has having 'declared a war of eternal opposition between men and women', and being responsible for a 'perverted account of personal relations and of social composition'. This underpins his argument and allows him to cast feminism as *against* men – which from our overall definition of feminism as seeking *equality* is again erroneous. It is in fact interesting to consider that Lyndon sees the attempt to achieve equal treatment as a declaration of 'war'. The idea that those who are used to privilege experience equality as oppression is perhaps useful here.

Lyndon's (1992) initial definition of feminism can therefore be seen as a caricature, which deliberately exaggerates in order to create the straw men and false dichotomies to follow. He also seems to fundamentally misunderstand the call for equality, as to seek an equal balance is the very opposite of seeking to overturn patriarchy and replace it with matriarchy, which is what Lyndon seems to have in mind. At the end of the prologue Lyndon says that to be anti-feminist is not to be anti-woman (p10) – but he doesn't seem to acknowledge that to be anti-patriarchal is not to be anti-man.

Having looked at all of this, see if you can write a short summary of the doubter's response.

TASK 5.6

Write a short paragraph (100–150 words) summarising the counterarguments to Lyndon's (1992) position.

Example: It is doubtful, overall, whether any of Lyndon's (1992) 'little list' of facts does indeed suggest that women are advantaged over men, and that society is therefore not patriarchal. While his list includes some important points regarding ways in which men seem to be treated poorly, he does not demonstrate that these add up to a meaningful advantaging of women. No definition of patriarchy suggests that it can be seen as beneficial to all men in all situations, and Lyndon's argument is based on this misconception, along with the idea that feminism seeks an opposition between the sexes rather than equality, which in fact entails seeing both sexes treated equally well. Recognising these flaws in his reasoning allows the points he makes to in fact be seen as feminist issues in themselves, as any attempt to achieve gender equality would also need to address these problems.

Having gone through the counterarguments and the counters to the counters, let us consider then how reading Lyndon's (1992) text has affected your own position on feminism.

Again, answer the question, are you a feminist? Why or why not? Has reading Lyndon (1992), and our response to it, changed or strengthened your position?

Before we leave Lyndon (1992), it is perhaps worth noting that my choice of this text could itself be seen as a straw man. Lyndon is a journalist, not an academic, and his argument is deliberately exaggerated in some ways for effect. If you wanted to counter my counter to Lyndon, one way you could do that would be to point this out and to argue that I have selectively taken one small section of his book in order to caricature his argument.

I would argue that this is not true, and that the treatment of his work here is fair. But then again, I would say that. I will leave you to use the strategies discussed here to decide on your own answer.

It is also worth noting that at the beginning of this chapter we said that counterarguments are not always adversarial. The deliberate choice of a text that was directly opposed to a chosen position, however, inevitably meant that the response constructed here was one of outright disagreement. In order to look at how we synthesise arguments and counterarguments to produce stronger overall positions, we need to think in more detail about how to structure arguments and incorporate different perspectives and positions, which brings us on to our next chapter.

Summary

- Counterarguments are a vital part of the academic process, whether to test existing positions, generate responses to them or explore and develop our own arguments.
- Looking for factual errors, logical fallacies or examples of faulty reasoning is an important first step in considering counterarguments.
- When considering an argument, it is important to think about it on all levels of abstraction – in other words, from the detailed level of individual pieces of evidence or facts, to the broader, more zoomed out perspective of approach and overarching question.
- All arguments are based on a particular selection of evidence, and thinking about what was deliberately *not* included is as important as thinking about what *was*.
- Think about how you are approaching the text as a reader. Try and *think like a believer* and *think like a doubter* and see how that affects your view on the argument.

- Consider possible alternative approaches, methods or interpretations. Looking at the paths not taken will help you to evaluate the path that was.
- Think about the purpose of any argument and the assumptions underlying it to see if this suggests any counterarguments. All arguments exist within particular contexts, and considering these contexts will help you to explore different approaches.
- Everything discussed here applies equally to your own arguments as it does to the arguments of others. It is as important to be critical and open to challenge when producing your own work as it is when approaching the work of others.

FURTHER READING

Chapter 4: How to criticise arguments from Bonnett, Alistair, *How to Argue*, 2nd edn (Pearson, Harlow, 2008).

Metcalfe, Mike, *Reading Critically at University* (Sage, London, 2008).

NOTE

1. For a more detailed exploration of this example, see Rush (2020).

6

Structuring arguments

When it comes to making arguments in an academic context, the form is almost as important as the content. Not only can the way we make our point have a huge amount of impact on how that point is received and understood, the way that we make our point fundamentally affects what we are saying in the first place.

How to structure an argument, whether it be in an essay, report, exam or presentation, is one of the most common student anxieties at all levels of study. Underlying this anxiety is often a sense that there is a secret to this that is being held just out of reach by their lecturers and the other students who seem to know what to do – that there is a magic recipe book somewhere behind the curtain and that opening it will unlock all the mysteries of structure and make producing an argument simply a matter of following a predetermined pattern.

There is no such recipe book, and this chapter is not going to attempt to provide a set of patterns that will provide you with a set of instructions for building your arguments. Instead, we will explore how the way that an argument is structured is actually a response to a set of key questions about what you want to achieve, and that rather than being a predetermined mould into which your content is poured, the structure and the argument are fundamentally interlinked and, to borrow from Marshall McLuhan, the medium *is* the message.

In other words, structure is not something secondary, structure *is* argument in a very important sense. To make this a bit clearer, let's start at the beginning.

Teaching an everyday skill

In order to think about how we structure arguments, let's start by thinking about something much more straightforward – or at least something more everyday.

I would like you to imagine that you have been asked to teach someone a basic skill or how to perform a simple task that you know how to do well, in five minutes. This can be anything – how to make a good cup of tea, how to hold a cricket bat, how to operate a mobile phone, or how to tie a sari – the important thing is that you know how to do it well, and that you feel like you could explain it to someone else in five minutes.

TASK 6.1

Having chosen the simple skill that you would like to teach someone, write (or think of) a brief step-by-step plan of how you would go about doing it. Remember that you only have five minutes.

If you can find someone to teach your skill, that would be ideal. Even if they already know how to do what it is you are trying to teach, it is still an interesting exercise to see how effective your plan would be in reality. If there is no one that you could do this with in person, then see if there is anyone you could message or email your instructions to and ask for feedback on how they found it.

If it's not possible for you to actually teach someone in reality, either in person or remotely, then take a moment to imagine teaching someone – that is, walk through the process in your mind.

Once you have done this, consider the following questions:

- Did it/would it work? Why/why not?
- What decisions did you have to make when you were preparing? What did you have to think about? Why?
- Could you have done things in a better order? Did you miss any stages out?
- Why did I ask you to do this?

Before considering the answer to any of these questions, let me first give you an example.

How to make a good cup of tea

Equipment: Kettle, mug, teaspoon.

Ingredients: Tea bag, splash of milk (semi-skimmed).

1. Fill the kettle with enough water for a mug and boil.
2. Put a small splash of milk in the mug – about half a centimetre should be enough.

3. Add the tea bag.
4. Pour in the recently boiled water and leave to brew for a minute or two.
5. If necessary, bring the bag to the surface with the teaspoon, then dunk the bag repeatedly until the tea is the desired strength.
6. Remove tea bag and throw away.
7. Drink.

Certain steps in this example could provoke strong reactions in some British readers for reasons that we will explore in a moment. But before we do that, let's consider the four questions from a moment ago together.

Did it/would it work? Why/why not?

This question is much more difficult to answer if you did not actually try to teach your skill to a real person but only imagined doing so. However, most of us have tried to teach someone something at some point in our lives, and what is often surprising or frustrating about the experience is how difficult it can be to do effectively. Instructions that seem straightforward are misunderstood, terms that seem obvious need unexpected levels of explanation, or you realise that you should have explained certain things in more detail before you even got started. You can also find yourself facing questions that you don't know how to answer – even with a topic you know well. In order to think about why this is, let's consider the second question.

What decisions did you have to make when preparing?

Whether consciously or unconsciously, the moment that you started to think about how to teach your simple task to someone else, you started to make a series of decisions. Some of these would have been obvious and seemed quite straightforward, such as what order do I have to do things in? Is there any equipment I'm going to need? What information has to be included, and what might be useful but can be left out? Is there anything I need to explain before I even begin?

Other decisions would have been much less obvious, and you might not have been conscious that you were making them. Many of these would have related to something that was not even explicitly mentioned in the instructions I gave you for the task – namely the person that you would be teaching.

If we had done this exercise in a classroom, you would have known who it was that you were going to teach, and you would have made decisions based on what you knew or could guess about that individual.

Again, either consciously or unconsciously, you would have made certain assumptions about them, which would have made you think differently about the best way to teach them. For example, their age, the country they were from, how much you shared a common language, their class background, gender and so on would all have told you something about what you presumed their existing level of knowledge of the topic at hand to be, and thus what you did and didn't have to include in your mini-lesson.

Even if you were not preparing this task for a real individual you would have still made certain assumptions about your imaginary audience – for example, what you thought needed explaining, what you didn't think needed explaining, what was 'common knowledge', and what sort of form would make sense to them and to the matter at hand. The only instruction I gave you was that you were to prepare to teach 'someone' and it is likely that you did not stop to consider what that meant and instead, broadly speaking, filled the position automatically with someone who looked a lot like you.

To illustrate this, let's look at the example I gave you. Everything that I chose to include was very Anglo-centric and relies upon a strong understanding of a British cultural context to make sense.

This is partly because tea is a very important part of British culture. This is not to say that all British people drink, or like, tea, but the vast majority of British people would be familiar with it and the way in which it is commonly drunk in the UK. Because this is the first assumption that is made here – this is a good cup of tea from a British perspective and there are lots of other ways of drinking tea from around the world that look nothing like this. This affects everything included here – I do not explain what a tea bag is, or what type of tea is inside it, although I do specify which type of milk (which is called something different in other countries); I do not explain how to use a kettle or what one is (almost all households in the UK have a kettle, which is very much not the case in other countries); I do nothing to explain when or why tea is drunk (as I drink it all day, all the time); and I do not explain certain key items of terminology, for example the verb 'brew'.

What all of this tells us is that even the obvious questions are in fact not obvious – you can't answer them without asking the more complex ones, and the fact that we often answer the complex ones without really thinking about it tells us something important. How obvious or logical you found my example above will depend largely upon how closely your cultural knowledge and assumptions match mine, as will the success of my attempt to teach you. That is because the other element that we needed to explicitly consider when undertaking this exercise, which was not mentioned in the instructions, was ourselves – the teacher. Because I did not consider my own assumptions before coming up with my set of instructions, I wrote something that would only have been properly intelligible to someone with certain pre-existing knowledge.

To think about this in a slightly different way, imagine that instead of asking you to teach someone a simple skill, I had asked you to tell them a story about a significant event from your childhood. It is perhaps easier here to think about everything that you would have had to explain before you could even start telling the story and about how key framing the context would have been. You do not consider your own history to be 'common knowledge' in the way you do the tasks you are familiar with, and so the order in which you told the story might also have required more thought as you considered what background you needed to include and how best you could convey the significance of the chosen event. But the principles would have been exactly the same – you would have had to consider what message you wanted to convey, what you needed to include to make sure that message got across and made your choices accordingly.

Could you have done things in a better order? Did you miss any stages out?

There are many reasons why another order might have been better, or you might have missed a stage of your explanation out, but outside of simple error, most of these would have related to the considerations above – namely the relationship between the assumptions you made about the individual you were teaching and that individual in reality.

For example, in the case of my 'good cup of tea' recipe, it might have made sense to include a brief paragraph explaining the place of tea in British culture and how and when it is drunk. This would have both served to help my pupil understand the instructions that followed and, vitally, it would also have helped them to understand why I chose that topic in the first place. In fact, this would have been useful even to someone who already had all of this knowledge – as it would have given them a context for what followed.

One thing that was definitely missing from my instructions was a consideration of other possible ways of making a cup of tea. There is a cliché that one of the most heated debates it is possible to have with a British person is whether you should put the milk in first or second (i.e. before or after the boiling water is added to the tea), and while this might seem like a joke, some people do take it very seriously. Likewise, there is no discussion over different types of tea here, whether that be loose-leaf or tea bag, green or black, brand name or variety (English breakfast? Assam? Darjeeling?).

This was a deliberate choice based on the constraints of the task. I only had five minutes with my pupil and the task was to show someone how to do something I knew how to do well. I therefore decided that I didn't have time to go into any of the many possible debates that could have been considered and these would anyway distract from the task at hand.

Why did I ask you to do this?

As you may have already realised, this is an exercise in structure. Explanatory or narrative structures, such as teaching someone how to do something or telling them a story about yourself, are so much a part of everyday life that we often do not realise that there is any deliberate act involved in building them. But there is, and the key insight to take from this is that structuring is an everyday skill that you already have, to a greater or lesser extent. Structuring an argument in an academic context is an extension of precisely the same activity and involves many of the same skills.

This task also demonstrates how practical considerations or constraints shape the final structure as much as the content of what you want to say. In this case, the five-minute time limit, and the fact that you were asked to teach an individual (as opposed to, say, a class full of people), determined much of the way in which you had to approach the task and what you could and couldn't include. It also wasn't the case that simply knowing about your task was enough to tell you how to teach someone how to do it well. In order to do that, a whole series of other questions had to be considered, and it was the answers to those questions combined with the knowledge that produced the final result.

Structuring an academic argument is simply a different version of the same process, and involves many of the same questions, albeit in slightly different forms. The structure and the argument are intertwined, and one does not exist without the other. Or to put it another way, making any argument is not just about what you want to say, it's about who you are saying it to, the way in which you want or have to say it, and perhaps most importantly of all, why you want to say it in the first place.

Steps in the process

In order to map the everyday skill onto the academic skill, let's consider then – what stages do we go through when constructing an argument?

TASK 6.2

Take notes on the stages that you go through (or think you *should* go through) when building an argument in an academic context. What do you have to consider?

As we have already touched upon, there is no one set recipe for putting an argument together, not least because the context or the circumstances in which individual arguments are produced is so important in determining the steps that need to be taken.

However, the following list represents a few important stages to consider, although they do not necessarily need to happen in the order listed here or to all happen on every occasion.

1. **Understanding your purpose** – This could involve analysing an essay question or assignment, or it could be about working out your response to the debate or issue at hand in a particular week of a module so that you can discuss it in class. Whatever the situation, you need to work out what it is you are trying to do – what is the context you are working within, and how does your argument relate to that? At this point, you will also have to think about the questions raised by the exercise above, namely, what is your audience? What assumptions can you make about them, and what does that tell you? Will your audience or reader have the same knowledge as you, or are you introducing something new to them, for example? Your understanding of what it is you are trying to achieve may change as you go through the process of building your argument, but this is the vital foundation of everything else.
2. **Reading, research and brainstorming** – Often, you will already have done a lot of the reading and research before you come to making, or considering, a particular argument. This is especially true when you come to write an essay, start a research project or sit down to tackle an exam paper, as you will have done a lot of reading and research during the normal process of studying a module. Before deciding on structure, however, you need to think of everything you know that might be relevant – all the ideas, thoughts, references, quotes, facts, debates and discussions that you have come across. This process is called *brainstorming*, and is a way of taking stock of what raw materials you have to work with. It is also a way of working out what gaps there are in your knowledge, and therefore what extra research and reading you need to do. This process can therefore be a circular one that takes some time, but it is a vitally important one.

3. **Decide what your overall point is** – It is often impossible to pinpoint precisely the moment at which you come up with your overall point – that is, the final conclusion that you want your argument to lead to, or the position you want to take on a question, or in a debate. Sometimes this is the first thing you determine, and sometimes you need to go through every stage here in order to work out what you think. However, whatever happens, it is important to make sure that you do go through the brainstorming process (point #2) before finally settling on your overall point, as otherwise you might go with your instinctive first response to a question, situation or text and ignore a lot of other possibilities. Equally, it is vitally important to take this step before moving on to finalising your overall structure – as how can you plot a route to your destination if you don't know what that destination is? Think back to the exercise at the start of the chapter – how would you have worked out how to structure your mini-lesson before you knew what you were trying to teach someone to do?
4. **Consider counterarguments** – Having come up with an overall point, you need to think about all the possible challenges to it (see Chapter 5). This is something that you should consider at every point in the process – it is important to be as critical of your own ideas as you are of others. As such, you will do this as part of the brainstorming process, you will do it while deciding what order your points will go in, and you will do this right up to the last possible moment. You might even do it when considering the very purpose of your argument – but what is key is that you are always open to questions, to possible alternatives and to challenging previously unconsidered assumptions. You can then use these to either reconsider your argument or reinforce your position by showing how it already takes into account and overcomes any opposition.
5. **Decide upon a skeleton** – A skeleton, in this context, means an outline, or very brief and simplified structure. You can do this by considering all the raw material that you have from step 2, and thinking about how it groups together or can be organised. You can also think about prioritising your points – that is, what is most important and what is useful but perhaps not essential? Doing these two things (i.e. considering groupings and what needs to be included and left out) can be done as a way of helping you to work out your overall answer as well, and rehearsing different possible outlines can be a good way of testing what you think and considering possible counterarguments. Practical considerations will also play a part here, as the word count of an essay or the length of a seminar will determine how much space you need to plan for. Another way of thinking about this is as sketching a rough outline of the picture you want to draw before you fill in all the detail – and it's much easier to make changes to a rough sketch than it is to a finished painting.

6. **Prepare a detailed structure** – In other words, flesh out your skeleton or colour in your sketch. If this is for a piece of writing, an assignment or a presentation, you need to have a proper plan and not just a list. This means that you should know not just the order that your points are going to come in, but how they link together, and how they all lead to and support the overall conclusion. You should know where each piece of evidence comes in, and how counter-arguments are going to be incorporated. You should know in detail exactly how every part joins together, and what steps you are going to take your audience on from the beginning of the journey to the destination. Doing this work at the planning stage will make the writing or preparation much easier, as you will already have done the thinking, and will not hit so many obstacles or run down as many blind alleys as you might do otherwise.

To illustrate all of the ways that this might work in practice, we're going to walk through a range of different questions and look at how each suggests different structures.

Example structures

As discussed above, there is no magic recipe book of structures and every discipline has its own conventions of how to structure different genres (see Chapter 2) and the arguments within them. Some of these are very explicit, such as Science Paper Format (SPF), others are more implicit with conventions that need to be learned and understood through experience. Understanding these is vital to successful study, but within all of these, there are also some basic forms that it is helpful to be aware of.

These basic forms are very much just a starting point. Each one, for example, could constitute the entirety of your argument, or they could form the structure of individual elements within that argument, and simply represent one building block in the overall structure. Often, they will need to be adapted to suit the constraints of the assignment, topic or specific institution, but they can be useful as a way of generating ideas and building your basic skeleton.

The logic chain

One of the most common forms of argument involves showing how one proposition leads on to the next, and how the whole series builds to a clear and logical conclusion. You can think about this as a chain, where each element is linked to the next with a logic word – and, but, so, therefore, etc.

For example (see the extract from Ngugi (1986) in Chapter 1):

> Under colonial British rule, Kenyan children were forbidden from speaking their own language, and forced to use English.
>
> ↓
>
> AND throughout the education system, up to university level, English language qualifications were more important than any others.
>
> ↓
>
> THIS SHOWS THAT language was seen as a vital part of the colonial project, and of controlling the colonised population.
>
> ↓
>
> THEREFORE, on becoming independent, it was as important to reclaim language and decolonise the mind as it was to reclaim land and decolonise the country.

The pillars and roof

In other cases, you could have multiple different propositions that all support the overall position. These propositions all lead to the final conclusion, but do not necessarily depend on one another. Think of this like a set of pillars holding up a roof, where the roof is your overall position and the pillars are the reasons supporting it.

For example (see Rush, 2020):

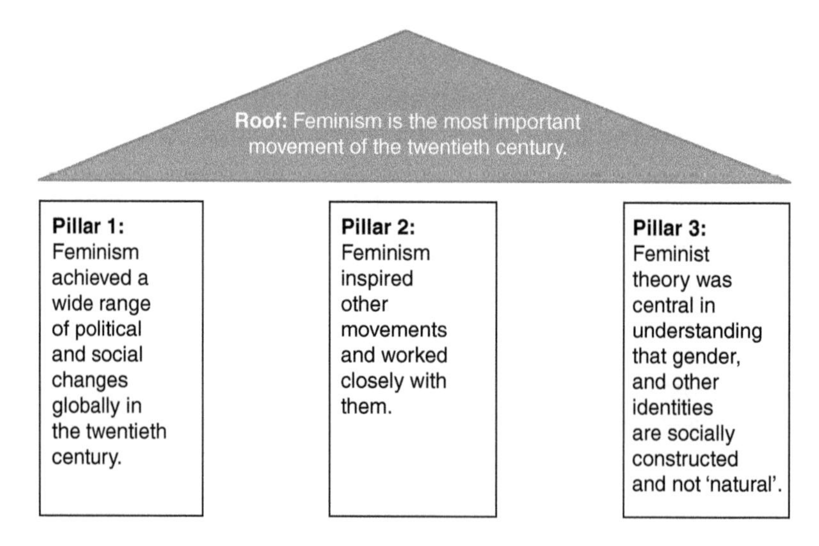

Figure 6.1: Pillars and roof diagram

The block

Sometimes there will be clearly demarcated sides to an argument, as, for example, when you are asked a yes or no question or whether you agree with a certain proposition. For example, the question, 'Should we buy clothes from shops that produce their goods in sweatshops?' suggests a structure whereby you have two 'blocks':

All the arguments that we should vs. all the arguments that we should not.

If we want to come down on one side or the other (that is, if we want to say broadly speaking yes or no) then only the two 'blocks' are needed, followed by a conclusion. More likely, however, is that we would come up with a synthesis of the two positions that drew on elements of each side to construct an overall position that brings them all together – often with some input of your own included that helps to make the final conclusion yours.

For example, consider Lynne Pettinger's (2012) blog piece that we looked at in Chapter 4, which we could use to help us build the following argument:

> We should stop buying goods in shops that produce them in sweatshops, as workers are exploited and endangered by the conditions they are made to work in, and as consumers, we can put pressure on brands to improve working conditions by denying them our custom.
>
> ↓
>
> BUT, stopping buying these goods will harm workers, as they will lose their jobs and be denied the freedom and choice that the work allows them, therefore we should continue to shop and exert pressure in other ways.
>
> ↓
>
> HOWEVER, casting exploitative work as a positive thing, simply because it gives choice to workers is wrong. A choice between dangerous and poorly paid work or nothing is not really a choice at all. Other means of exerting pressure on brands will be less effective than simply not giving them our money, and to continue to shop while claiming that this is helping workers is ultimately a self-serving position that allows us to justify our desire to buy cheap goods and continue to feel good about ourselves.
>
> ↓
>
> CONCLUSION: Both choices are imperfect and will cause some harm, but we should not continue to buy goods from companies that produce them in sweat shops as otherwise we will continue to force people to work in exploitative conditions.

Note how the inclusion of counterarguments and the rebuttal of these arguments has served to make the final conclusion much stronger. Note also how each stage of the argument builds on the one before, so that the final effect depends not just on the points that have been made, but how those points have been ordered.

Even if you were simply choosing one side of a debate over another, you would still need to consider carefully what order you put your points in within each block, in order to most clearly make your argument. For example, look back at the Donald Trump example (Chapter 4, p67) where I told you that my explanation was an example of how to structure an argument for effect.

The question there was – did Donald Trump really say that Republican voters were stupid, as a meme claimed he did? An outline of my answer to this question could be laid out something like this:

> YES:
>
> A – the quote given in the meme sounds like Trump (it matches his rhetorical style) and sounds like other things that he would say.
>
> B – the meme includes a reference, which suggests that we can verify this as we know who wrote it.
>
> HOWEVER:
>
> NO:
>
> B – the reference is incomplete, and the picture attached is also unreliable.
>
> A – the fact that this sounds like Trump plays into our desire for this to be true – in other words it uses confirmation bias to its advantage.
>
> B – the reference that we should be interested in is in fact the information on who made the meme – the Other 98. Noting this suggests a clear motivation for producing the meme.
>
> A – although the quote *sounds* like Trump, it in fact includes factual errors, which show that it is written far later than it claims to have been.
>
> CONCLUSION: Trump did not say this.

Note how the points in the second section mirror the order that they occur in the first, so that the argument pivots around the question of verifiability and the structure suggests a sense of balance and completeness.

Point-by-point

Rather than simply grouping your points in contrasting 'blocks', you can also organise them thematically, according to the area they relate to. In some cases, this is an obvious route suggested by a question (e.g. 'What factors affected the development of the political idea of a unified Europe throughout the twentieth century?') where your answer could be organised around economic factors, cultural factors and political factors). It can also, however, be used to deal with a yes/no or for/against type question, as an alternative to the block structure.

For example, consider the statement – 'Feminism is wrong to say that society is patriarchal'. Rather than simply grouping all the arguments for this statement and then following up with all the arguments against it, which would be a perfectly reasonable way of structuring a response, we could use Neil Lyndon's (1995) text, which we looked at in Chapter 5 to approach this in the following way.

> Argument – society cannot be patriarchal if women are advantaged over men in any way.

THEME 1 – Children:

> Y: Unmarried men – no rights regarding their children. Men also have no rights in cases of abortion.
>
> N: Law has been changed to give unmarried men rights; argument about abortion needs to be considered carefully in relation to historical context; in both cases, we do not know whether they are serious issues.

THEME 2 – Marriage:

> Y: Divorce courts routinely 'strip men of their property and income'; widowers do not get same benefits as widows.
>
> N: Benefits discrepancy is unfair, but is a result of assumptions about gendered economic roles, which are themselves patriarchal; 'property and income' of marriage is jointly owned – Lyndon's description is itself patriarchal.

THEME 3 – Financial rights:

> Y: Men cannot be classed as dependents for benefits; women get to retire earlier than men.

N: Female retirement age being changed; ideas of who can be a dependent again structured by patriarchally gendered economic roles.

CONCLUSION: There is no evidence that women are advantaged over men, and even if they were, the argument is fundamentally undermined by the flawed definition of patriarchy being used.

Note how the arguments for and against are split here across three themed sections, with the ordering of points within those sections again important in terms of the overall effect produced.

Competing readings

In an academic context, it is very common for any argument to involve not a straight fight between two sides, or a straight choice between yes and no, for or against. Instead, what you are often being asked to do is to think about multiple different ways of looking at a question or issue.

For example, consider the question that we looked at in Chapter 3 – 'What is literature?'. To discuss this, we used some extracts from Terry Eagleton's essay of the same name, where he does precisely this – describing different ways of answering this question one after the other, in order to demonstrate that literature is not an objective description of what is beautiful or exceptional in writing, but rather a product of social ideologies and power structures.

Or for an argument grounded in something a little more popular, consider the following:

Frozen is no different from a traditional Disney film because it is about princesses and features a heterosexual love story.

↓

BUT – The love story at the heart of the film is in fact about the love between sisters, and female independence, and heterosexual relationships are actually an obstacle to rather than necessary for the protagonists' self-actualisation. *Frozen* is therefore a feminist story at its heart.

↓

IN FACT – It goes even further than this, as the need to break out of defined and toxic social gender roles (see, for example, the song 'Let It Go') is pivotal to the story, and so it can be seen as not just a feminist story, but an allegory for other LGBTQI+ experiences.

All of the examples given here are perhaps easiest to imagine being applied to an academic essay. However, they are useful for any context in which an argument is made, whether that be in a presentation, in a classroom discussion, when making a research proposal or when interpreting data. Arguments are everywhere in academia, and it is important to look for them even when they are not explicitly mentioned. Setting a research question, for example, involves making an argument about why this question is important, or how it helps to fill a gap in the literature (i.e. something that no one else has looked into). Interpreting data involves making an argument about whether (and how) a research question has been answered, or whether a hypothesis has been proven. In each case, a structure is needed, and these basic structures can be used as a starting point.

It is also important to remember to look out for structures like these when you are reading. Keeping your eyes open for effective ways of building arguments in the work of others is vital in improving both your own ability to structure, and your understanding of how arguments are made in your subject.

Same question, different structures

In order to think about how you get from these basic skeletons to your final argument, we are not, here, going to go through the rest of the process of planning an argument in detail, whether that be for an essay, a presentation or a report. We have already looked at the importance of knowing exactly what order your points will come in, how they link together and how to use evidence to support those points. We have discussed the need for you, as a student, to learn the conventions of your discipline and its genres (see Chapter 2). We have also looked, in Chapter 5, at how to generate and evaluate counterarguments.

Instead, here we are going to think about how the same question or issue can lead to many different structures depending on the purpose and context in which that question is being asked.

Consider the following assignment questions, and think about how you might go about structuring your response to them. Is a structure suggested by the question? Is there more than one possible way of structuring a response here? Use the example forms given above to help you.

1. Feminism is the most important movement of the twentieth century: discuss.
2. Discuss representations of identity and belonging in postcolonial literature using *two* novels from the course.
3. What caused the collapse of the British Empire in the 1960s?

Now let's look at each question in turn and discuss which possible structures might work for our response.

1. Feminism is the most important movement of the twentieth century: discuss

As this seems to be a classic for/against question, the obvious structure to use here might well seem to be a block, with all the arguments that feminism is the most important movement, all the arguments that it is not, followed by a synthesis. If you chose this route, you might well argue that another movement deserves the title of 'most important movement'.

This question was, in fact, written by me, on a module that I used to teach, and this structure would very much not have matched my expectations as a marker. On that module, the lecture I gave on this topic explicitly made the argument that feminism was the most important movement of the twentieth century, and laid out why. It never occurred to me, therefore, that students would choose to dismiss feminism and instead argue for something else (e.g. trade unionism) instead.

The question that I was in fact asking was something more like, 'what arguments can be made that feminism is the most important movement of the twentieth century and are they valid?' and the response I was expecting was something more like the 'pillars and roof' example given above, where multiple different arguments about this proposition were explored, with counterarguments integrated into each section. Any student response could conclude that my position was incorrect (that feminism was *not* the most important) but in order to get a good mark, they had to at least consider the arguments I had made.

Now, you could argue that I should have written a better question, or at least one that more closely matched what I wanted students to do, and in many ways you would be absolutely right. However, I assumed that it would be obvious what I meant based on my lecture, and did not think that students would ignore that when answering this question (although some did). I can also most certainly guarantee that I am not the only lecturer who writes poorly worded questions.

The key here is to remember what we discussed at the start of this chapter, namely, what are the assumptions that underpin the question? This also relates to the idea of hierarchies of questions that we looked at in Chapter 1, where one question suggests a number of others, and the trigger questions we looked at in Chapter 4 – most importantly, 'why did my lecturer set *this* question? Why this one and not another?'.

Questions at university always exist within a context, and when we respond to them, it is vitally important that we think about that context as clearly as we can to ensure that we are building the most effective argument.

2. Discuss representations of identity and belonging in postcolonial literature using two novels from the course

This is a very open type of question, which some students hate and others love. Often, lecturers set questions like this because they are trying to be helpful – they want to give you licence to determine which direction you will take the argument in. Questions like this are also useful practice for thinking about how you might end up approaching self-directed research – that is, starting out with a broad area of interest within which you have to find your specific question.

However, these questions can end up becoming a trap if they lead you to discuss issues in a way that is too broad or vague. The key here is to establish a more focused context – what part of this broad area do you want to talk about? Again, you can think about this in terms of the trigger questions in previous chapters, and in particular that same question – why has my lecturer set this? What discussions, debates, lectures or texts did we cover that relate to this, and what clues can they give me about what I should be discussing?

Another trigger that can be useful in responding to questions like this is – what is it about this topic that interests me? The implied question here is – what do *you* want to say about belonging and identity in postcolonial literature? In order to answer that, it can be useful to actually set yourself a question, or at least make sure that you have a very clear focus.

From either starting point, lots of different structures are possible. You could construct a block structure here – e.g. by showing that your chosen novels give two different representations of identity and belonging, or by arguing about whether, for example, the representations in question are fundamentally positive or negative. You could just as easily have a 'competing readings' structure, where a surface reading of the novels suggests one interpretation, but a more complex examination suggests something else entirely. You could argue that the novels are in fact not about identity and belonging at all, and that these themes are projected onto them only because they are postcolonial.

Problematising the question in this way can be a really useful way of constructing an original or critical response to the question. By showing that there are unconsidered assumptions in the question itself, you demonstrate not only your knowledge, but your understanding of the context and debates within which that knowledge exists. Or to put it more simply, an interesting argument is often as much about an interesting approach to the question as it is strong knowledge of the subject at hand (we will explore this further in Chapter 9).

What is important is that the approach that you take to the implied question will determine the type of structure that will work – and to remember that what you cover in your argument needs to relate explicitly to what you learnt in the module you are building the argument for.

3. What caused the collapse of the British Empire in the 1960s?

The obvious suggested structure here would seem to be the 'point by point', where you could look at economic factors, political factors and socio-cultural factors (for example) in separate sections, and then make a case for which is most important.

However, you could just as easily do a hybrid of the 'chain of logic' and the 'competing readings' here, where you argue the following:

> It was the changing social and cultural conditions that we can see flourishing in the 1960s, and which were created by the development of a globalised economy, that meant that Empire was no longer acceptable politically
>
> ↓
>
> BUT – These changing conditions were in fact a symptom, not a cause, and the rise of the USSR and US as superpowers underpinned these developments. This is what changed the global political climate, and therefore social and cultural conditions, making Empire untenable.
>
> ↓
>
> BUT – All of the above are fundamentally linked to the Second World War, which bankrupted the UK and left it in economic and political debt to others. This is the root cause of the rise of the USA and USSR as superpowers, of the changing political and global cultural climate, and so on. Thus, the Second World War is the single most important factor in the collapse of the British Empire, even if it took until the 1960s for it to fully play out.

Note how the structure of the argument makes the conclusion more powerful. At no point are any of the points made proved untrue, they are rather shown to be symptoms of something deeper – and the argument is made more powerful by keeping this final cause in reserve.

There's a certain rhetorical flourish to this, whereby each stage convinces the reader but then says, 'aha, you thought all those were important, but they were in fact caused by this other thing, which is even more important!'. The impact of this argument would be much altered if you simply started by saying – 'the collapse of the British Empire was brought about by the Second World War, which caused the following ...'. The content would be the same, but the effect on the reader would not.

In each example here, it is clear that there are multiple different structures that are possible for each question, and that what is key is to think about what question you are actually answering and what other questions you need to think about before building your argument.

What you then need to consider is how to structure an argument that will produce the desired effect on your reader, and to do that we need to move on to thinking about style, rhetoric and how we say what we say.

Summary

- Structure is a key student anxiety, but it is in fact a skill that we all already possess to a greater or lesser extent.
- Building an argument at university involves the same processes that we go through, often unconsciously, in our everyday lives when we explain ideas or tasks to people or tell a story.
- In other words, the academic skill is the same as the life skill – just in a different context.
- Structuring an argument is not secondary, but is in fact as vital as what you want to say. *How* you argue has as much effect as *what* you argue.
- There is no magic recipe book of structures, and how you structure your argument should be based on what you want to achieve, on the context that you are working within and your intended audience.
- There are some simple forms that can be used as a basis for structuring an argument, but these are basic building blocks that need to be adapted to the task at hand and to the conventions and constraints of specific disciplines and institutions, and not predetermined recipes to be uncritically applied.
- Questions at university always exist within a context – whether that be the context of the discipline, the particular module or even the class group within which a discussion is taking place. When we build arguments in response to these questions, it is vitally important therefore that we think about and understand that context as clearly as we can.

FURTHER READING

Chapter 2: Structuring your argument from Bonnett, Alistair, *How to Argue*, 2nd edn (Pearson, Harlow, 2008).

Rush, David, *Build Your Argument* (Sage, London, 2021).

7

Writing an argument

Universities are broken up into departments not just to divide up the study of all the knowledge in the world into more manageable chunks, although that is of course part of the reason. Every discipline in academic study is in fact a set of arguments about the world – what it is, why it is and how it is – that are not so much a reflection of one bounded part of reality, but rather a way of understanding, and therefore constituting that reality. In other words, these disciplines are a set of answers to questions like – what happened in the past and why did it happen that way? Why do people behave the way they do? What's the best way of running a society and an economy? Why do we tell stories in the way that we do and what does that tell us about ourselves? How do our bodies work, and what is consciousness? And so on ...

To answer these questions, each discipline builds a model of reality, which as Stephen Hawking says, 'exists only in our minds and does not have any other reality (whatever that might mean)' (Hawking, 1988, p9). That model is shaped by a set of assumptions and worldviews, and when producing an argument in any given discipline, we are producing it in the context of those particular assumptions, worldviews and the set of questions that make them up (see Chapter 3). Those questions, in other words, make us a particular set of eyeglasses that we use to look at the world, and which shape both what we want to say in our arguments and how we need to say it – both so that we can say what we want to, but also so that it makes sense, and engages in conversation with the other arguments produced by the academic community that makes up our subject.

As such, in order to make an argument at university, we need to understand our content – what we want to say – but we also need to understand our context and it is that context which determines what language we need to use in different situations to make ourselves understood as clearly and effectively as possible.

To help us think about this, let's start at the beginning, with a question: what makes good academic writing?

No matter how little experience or knowledge you have of academic conventions, it is likely that you have a sense of some of the ways in which academic writing is different from other forms of writing – for example, from journalism, or writing a message to a friend, or writing fiction – and things that you should and shouldn't do when producing it. These could be really simple ideas – for example, that academic writing is more formal than other types of writing – or it could be more complicated, abstract ideas – for example that academic writing is critical and analytical.

TASK 7.1

Think about, and take notes on, everything you know about what makes good academic writing. What qualities does it have? What are the dos and don'ts? What questions do you have?

Here are some examples of things that you could have thought about, although this list is intended to be indicative and not exhaustive.

Formality – Academic writing is more formal than other types of writing and doesn't use slang or colloquialisms.

Evidenced and **referenced** – Academic writing is supported by evidence, and is clearly referenced so that the reader knows exactly where that evidence has come from.

Precision – Academic writing uses words very carefully in order to say exactly what it means. This includes using subject-specific language where appropriate.

Concision – Academic writing is concise, which doesn't just mean being short or brief, but rather means giving all the necessary information in the shortest amount of space possible. This is often determined by the word limit of a particular form or assignment, rather than how much can be said on a topic – which is usually somewhere approaching infinite.

Objectivity – Academic writing is objective, not subjective. That is, it is impersonal and not affected by emotion, opinion or belief, but is rather based on fact, logic and rational thought.

Language – Academic writing is grammatically correct, contains no spelling or punctuation mistakes and is well structured so that it flows logically and coherently.

Based on this, take a look at the following examples and see if you can identify any problems in terms of style:

1. People who smoke have health problems.
2. If the government introduces this policy, it will have a negative impact on the economy.
3. Most people would agree that Dickens is one of the most important novelists of the nineteenth century.
4. Smith (2019) honestly believes that childhood trauma explains all bad adult behaviour.

ANSWERS

1. 'People who smoke' = smokers. You could also argue that this is overly certain, as not all smokers have health problems, therefore 'smokers are more likely to have health problems' would be better.
2. As above, the statement is too certain. 'Will' suggests that the policy is definitely going to have a particular effect, but you cannot know this, as its results are in the future, and the over-certainty makes it easy to argue against. 'It is very likely to have a negative impact' would be better, as it expresses the same idea, and is not vulnerable to easy disproof, or accusations of overstatement. Language like this is often called *hedging*, which just means thinking very carefully about how strongly you want to make a statement – for example, on a scale from will, to might, to won't, or correct, to convincing, to absolute rubbish.
3. Who are 'most people'? If you are talking about literary critics, then say so, and reference the ones that you are citing. If you simply mean that it is a generally held view that something is the case, then you do not need this equivocal prefix – simply say it – 'Dickens is one of the most important novelists of the nineteenth century'. If most people really would say it, then it is likely that no one will argue with you.
4. 'Honestly' adds an emotive element to this claim and is used to imply that Smith's position is stupid or ridiculous in some way. Claims like this should be made more dispassionately, for example – 'Smith's argument (2019) that all bad adult behaviour is caused by childhood trauma is questionable'. You could also argue that 'bad' is a bit informal here, and so 'negative' might be better. It is fine to disagree with sources that you refer to, but do so in a respectful manner – so, 'Jones' argument is flawed' and not 'Jones' argument is stupid', for example.

There are many more things not included in the list above that could easily have been, including the use of the third and not the first person, how the work is formatted or the fact that academic writing is intended for a specialist and not a general audience.

All of these things are correct – some in all contexts, others in many. However, learning a list of rules like this, while a useful starting point, in other ways is not very helpful. This is mainly because any blanket instruction like this automatically leads to two questions. Firstly, there is the question of what the instruction means – what does it mean to be formal and is that the same in every context? How do I know which words are allowed and which aren't, especially if English isn't my first language? Where is the line between opinion and evidence? What sort of evidence is appropriate and what isn't?

Secondly, and perhaps more importantly, is the question of why. Why do I have to write in this way? Why do I have to learn a different language for an academic context? Why can't I just speak or write like I normally do? Why are contractions not allowed and why aren't I allowed to say 'I'?

In order to answer these questions, and to think about where 'rules' like this come from, let's look at some broader questions. In doing so, we will also see that many of the 'rules' we are taught are not in fact rules for all situations at all, and part of the problem with learning them in this way is that we don't know how to work out when they do and don't apply.

Clear and simple as your argument

Francis-Noel Thomas and Mark Turner's (1994) *Clear and Simple as the Truth* is a famous book on how to write well. In it, they are particularly concerned with the idea of 'classic style,[1] which we are not interested in here, and with how in order to produce good writing, 'writers must work through intellectual issues, not merely acquire mechanical techniques' (1994, p3), which we are.

They argue that rather than just memorising a set of rules, 'learning to write' is in fact concerned with, 'learning styles of writing, and ... styles derive from conceptual stands' (p6). These 'conceptual stands' are analogous to the sets of arguments that make up different university disciplines, which we discussed at the start of this chapter, and while Thomas and Turner (1994) are concerned with the specific conceptual stand that underpins their chosen 'classic' style, we can adapt some of their concerns to pose ourselves a set of questions which will allow us to respond to any given situation that we find ourselves in, and work out how to write or speak according to that situation and our purpose within it.

There are many different styles, and many different academic styles, across and within disciplines. Each is based on a set of principles and assumptions, a set of answers to certain key questions. The mechanics of style varies from discipline to discipline and from situation to situation, therefore, but the underlying principles are the same.

The important thing is to be aware of how the choices you make about the language that you use will affect your overall argument and why you are making those choices.

TASK 7.2

To think about this in more detail, take a look at, and take notes on, the following questions:

1. What is the purpose of academic writing? What is it for?
2. What does your discipline (i.e. the one that you study or are going to study) mean by 'truth'?
3. Who is your reader?
4. What is your relationship to that reader?
5. What does it mean to give your opinion in academic writing?

1. What is the purpose of academic writing? What is it for?

As a student, it is very easy to get lost in the treadmill of work and assignments, and even to feel like the main reason you are asked to do so much writing is because your lecturers want to make you suffer. It's important, however, to think about the bigger picture, and about why you are asked to do the tasks you are, whether that be to write an essay or a lab report or a reflective portfolio.

There are two key aspects to think about here.

The first is the relationship between academic writing and academic activity as a whole. In a very important sense, academic writing is not part of any given discipline of study, it *is* that discipline. As we discussed in Chapter 1, one of the key purposes of university is to produce and disseminate knowledge, and that knowledge has no existence outside of the writing (using the term in its broadest sense) produced by academics. Any act of academic writing is an engagement with and a contribution to that accumulated body of knowledge.

This applies to the writing that you as a student produce as much as it does to the work your lecturers do. That might sound like an intimidating prospect, but it is important to realise that a university student – whether undergraduate or postgraduate – is regarded as a member of the academic community, and their work is treated as such. Being a student at university is a very different role to being a student at school, as we shall explore further in a moment.

The second aspect, however, relates to the more practical way in which being a student is important in terms of the academic writing (and arguments) you produce. That is, you, as a student, need to demonstrate what you have learnt from your studies so that your lecturers can award you whichever qualification you are studying for. That means that whenever you produce an argument at university, you need to be thinking about what it is you are demonstrating in terms of the knowledge and skills you have gained from your studies in the particular module or context concerned.

Both of these aspects provide a very useful way for you to think about your purpose whenever you sit down to write. Different genres, for example, serve different purposes in academic discourse – see Chapter 2 – and are making a particular type of contribution to the wider disciplinary conversation. An essay, for example, is an attempt to construct an argument in response to a question, while a lab report is an account of a given piece of research that is delivered to both communicate the results of that research and allow others to repeat it.

Equally, it is vital that you remember the most direct purpose of everything you produce as a student, namely engaging with your most local academic community – your lecturers and fellow students.

2. What does your discipline mean by truth?

This might sound like an odd question, as surely there is only one type of truth? Another way to think about this question that might make it easier, then, is – what does a 'good' answer look like in your discipline? What does it mean to be 'right'?

One way of answering these questions is to think about the different purposes that your arguments might be serving. Are you explaining what, or how, or why? Are you giving the solution to a problem? Are you interpreting what something means? Are you analysing something scientifically? What does that mean? And so on.

Another way to think about this is to consider what *evidence* you are using to support your arguments. If you are studying data science, the answer will be very different than if you are studying literature, and in each case the nature of the 'truth' you are seeking will affect the language you can use to speak it.

As we saw in Chapter 2, what is given as a 'good argument', or what is said by a discipline to be 'true' is often not just about data, or facts, but rather a claim about what such data or facts can be said to mean and the logic that is used to get from one to the other.

All of this affects the content of what is said, but it also affects the style in which it is said. Different subjects, for example, would place different levels of emphasis on the items listed above. Sociology is less concerned with objectivity than psychology, for example, because it is interested in accessing and including individual and personal experience while psychology is concerned with maintaining a position as a science. Scientific writing is also very concerned with precision and does not like ambiguity, whereas ambiguity in philosophy and literature is (arguably) more tolerated. This is not (or not only) a matter of disciplinary taste. It is a matter of the types of question a subject deals with, and the types of answers it tries to find to them.

To return to an example used earlier in this book, consider how different the answer might be to the question 'what is beauty?' if you asked a biologist as opposed to a philosopher. While the biologist might talk about data collected on individual responses to different facial or body types and the neurological (or even evolutionary) functions attached to interpreting sensory data as attractive or repulsive, the philosopher might be more concerned with interpreting and contrasting different conceptual and aesthetic theories of beauty and how they have developed over time. Both are equally valid responses to the question, and are simply drawing on a different standard of truth or a different epistemological framework, depending on how you want to think about it.

3. Who is your reader?

In many situations at university, you will know exactly who your reader is – it will be your lecturer or class tutor who marks your work. However, you don't write as if this is the case, as that would be very odd – not least because your lecturer is likely to know far more about any given topic than you do.

If we don't write for a given individual, then, who do we write for? The answer is that we write for an imaginary or hypothetical reader and we ascribe certain characteristics to that reader.

Consider the following questions about the imagined reader for a piece of academic writing and see what you think:

- Are they intelligent?
- Do they know anything about the subject?
- Are they interested in what you are writing about?

The answer to all of these questions is yes. You can assume that the person reading your work is intelligent and that they have at least some working knowledge of your subject. Academic texts are specialised and specialist, and while the exact level of specialisation depends on the particular text (for example a journal article is likely to be more esoteric and require more pre-existing knowledge than an introductory textbook), academic writing is intended for an audience familiar with and engaged in the study of a particular subject, not for newcomers to it.

Practically, that tells us a few things about what choices we need to make. Firstly, it tells us that we don't need to include basic background information – there is a level of existing knowledge that can be assumed, including, for example, of terminology and concepts. It also means that we don't need to persuade or seduce the reader – this is not journalism where the reader needs to be 'hooked', and the reader can instead be trusted to deal with whatever complexity and difficulty is necessary to understand a particular topic.

4. What is your relationship to that reader?

If our reader is imaginary, then what is our relationship to them? In order to think about this, it is perhaps better to change our question and ask instead – who are *you*? As a writer, you do not write as yourself, as this would be as odd as writing directly for your lecturer.

Instead, as an academic writer, you are playing a role – an imaginary part, in other words – and you choose your language accordingly. This might sound strange, but it is actually something that you are already accustomed to doing and something that you do every single day.

The language that you use to talk to your friends is different from the language you use to talk to your parents, or to a shopkeeper, or to a policeman, or to your university lecturer. In any given situation, you automatically ask a series of questions about the context you are in and what you want to get out of it, and adjust your language accordingly. Writing, or speaking, in an academic context is no different.

So, what can we say about you, the student, as a writer? What characteristics can we ascribe to the imaginary role you are playing? Are you intelligent? Yes. You might not feel like it, but you have to act like you are. Are you capable – by which I mean, are you able to answer the question set and is your voice worth listening to? Again, the answer is yes, even if you don't feel that way. Are you inferior to your reader? No. Are you superior to them? No. Academic writing is a conversation between equals and it is written accordingly. This last can be a culture shock for people who grew up in an educational culture where they are expected to show deference to their teachers, but in a 'Western' university context, as a student you are expected to act as an (admittedly junior) member of an academic community of equals.

All of this informs our choices as producers of academic writing. It also helps to explain all the things on our list of 'rules'. We write formally as this is a professional and not a personal context, where we do not know the person to whom we are talking. Our work is referenced because we are engaging in a field where much has already been said, and evidencing what we say both gives due recognition to the work that has informed ours and allows others to understand the context that our arguments are made in. We are precise because the use of words is important and can have a crucial effect on our meaning – and so on. But these are choices that are made as a result of the role we are playing, the audience we are writing for, and what we want to achieve, not because there is an agreed rulebook of academic style somewhere that decrees what is right and wrong.

5. What does it mean to give your opinion in academic writing?

It should be obvious from all that we have said so far that 'opinion' has no place in academic writing. 'Opinion' as a general rule is personal and subjective, and does not necessarily have to be based on reason or fact, and you can't argue with someone's opinion or belief. Despite this, you will often be told that you need to give your opinion in academic writing and may even see the word used in writing guidance from your lecturers. Why is this?

The key here is that while you are not asked to give your opinion on a question when you make an academic argument, you are asked to give your *answer*. The key word here is that it is *your* answer. You are expected to give your response, and while that response must be based (as much as possible) on critical and objective analysis, what is important is that *you* are adding something yourself and not simply repeating or recycling existing information. This is what lecturers mean when they say that they want your opinion – they want to know what your position is, and as long as you are careful to always translate 'opinion' to 'answer' in your own mind, you should be fine. Ironically, given everything we have just discussed, this is an example of lecturers not being precise enough in their choice of words.

We will look at how exactly you go about doing this in Chapter 9.

Having thought about all of the above, look at the following examples, taken from real students' writing, and see if you can identify any ways in which they could be improved in terms of their language or style.

1. Of course, it's important, at this stage, to state that the features of syntax are not consistent across all languages. That's not suggesting that syntax doesn't exist across all languages, because it does, it merely states that each syntax is different.
2. Yet mainstream feminism does put more of their effort towards expanding the equality of females than males as females face more inequality.
3. At every moment throughout an individual's life, humans need to process a wide range of information that is constantly coming at them. Visual selection is a useful tool that enables individuals to distinguish between stimuli that is helpful for the current aim and stimuli that is not helpful at all.
4. Diving beneath the waves in the warm waters of the equator you may be met with a fantastical view of a beautiful and unimaginably diverse marine city, inhabitants of which swim with little to no effort through their colourful metropolis under the surface of the deep blue sea.
5. Comparatively, Chesler takes a different approach when pondering how necessary feminism is in the 21st century. Chesler is less optimistic and arguably more realistic when addressing this question.

ANSWERS

1. The problem here is *concision*. The second sentence is *redundant* – that is, it adds nothing and so is not needed – as all it does it repeat the first sentence in a different way. This is a waste of words, and also creates a vague and rambling tone in a discipline (linguistics) where it is important to be precise and clear. Linguistics as a subject is about explaining the workings of language, and in order to do this, it's very important to say exactly what you mean.
   a. Corrected version: 'It is important to note that the features of syntax are not consistent across all languages.' (16 words rather than 42.)
2. There are a number of issues here. The main issue is that it's difficult to understand exactly what the student is trying to say. There are grammatical issues ('mainstream feminism' is singular, and so it should be 'its effort',

not 'their effort' for example), and there is also something odd about the use of 'females' and 'males' – as if the student is not sure what terminology is acceptable in this context. Perhaps the biggest problem, however, is the metaphor that has been used to discussed equality. Can equality be 'expanded'? This is an odd word to use, and whilst it is probably reflecting nothing more than a struggle to express an idea clearly, it suggests to the reader that the student doesn't really understand what they are talking about. Perhaps, here, the student is too worried about trying to sound 'academic' when they should just be trying to communicate their idea as simply as possible.

   a. Corrected version: 'Mainstream feminism focuses more on issues affecting women, as they face the most problems with inequality.'

3. Again, the problem here is *concision*. This example is taken from an essay in the sciences and as such needs to be as concise and precise as possible. Significant improvements could be made here just by deleting words that are not needed. The first sentence, for example, is unnecessary context-setting and does not say anything useful or interesting – it is simply stating the obvious. The final clause is likewise redundant. It is often a useful exercise to do precisely this – look through your work and see what can be deleted without changing the meaning of what you say. This approach can be especially useful for novice writers in the sciences while you work on understanding what a scientific voice sounds like. Also note the grammatical issue – stimuli is plural and so the verb should be in agreement.
   a. Corrected version: 'Visual selection is a useful tool that enables individuals to distinguish between stimuli that *are* helpful and stimuli that *are* not.' (21 words instead of 51).
4. The evaluation of this example depends very much on which subject the student was writing for. For a module in marketing or tourism, this extract could be absolutely fine – if a little over the top. However, this example actually comes from the introduction to an essay by a marine biology student on the issues affecting coral reefs, and as such it is in completely the wrong style. The lecturer in this case commented, 'write like a scientist, not a journalist'. By that, they meant that a scientist's job is to explain what is happening and why, and not to present attractive descriptions or capture a reader's interest.
   a. Corrected version: delete, and move straight on to the relevant issues.
5. As with both 1 and 3, the issue here is one of concision, but in a slightly different way. In any subject, but particularly in the social sciences and humanities, referring to the perspectives of a range of different commentators is a vital skill, and referring to what has been said in different sources needs to be done concisely, otherwise you can spend an awful lot of words saying very little. As with example 3, think here about what words you can delete. The first sentence is redundant, as we clearly already know the question under discussion, and the second sentence tells us that Chesler is taking a different position, so we do not need to say it again.
   a. Corrected version: 'Chesler is less optimistic, and arguably more realistic.'

Having looked at the principles underpinning good academic writing, and corrected a few examples, let's consider a few more issues to do with language and style to round up some of the remaining questions that are often raised by students.

Can I use the first person?

One of the most common rules that students have drummed into them, whether in school or by enthusiastic English language teachers, is that you should not use the first person (that is 'I') in academic writing. The only problem is that this is not true, and often leads to much confusion for university students.

In actual fact, on many occasions it is absolutely fine to use the first person, while in others it is best avoided. What is important to understand is why it might or might not be appropriate.

Consider the following two examples. Do you think it is appropriate to use the first person in each case? Why/why not?

1. In this essay, I will argue that Shakespeare's status is a result of how he has been represented in British culture rather than any innate quality of his work.
2. In conclusion, I believe that criticism of Donald Trump is motivated by personal dislike more than a fair evaluation of his political performance.

One of the most common reasons given for telling students that they can't use the first person is that it sounds informal and subjective, as opposed to formal and objective. In terms of formality, this is not really true, at least not any more, and while it can have some truth in terms of objectivity, it is not really the use of 'I' that is the problem but rather what is being said by speaking in that way.

In example 1 above, there is no problem at all with using the first person – it is your essay, and that is what you are going to argue, and you have not undermined either the formality or the objectivity of the piece by saying it in this way.

The second example does feel more subjective, but the problem here is not the 'I', but rather the word believe. This sounds very much like something based on a personal opinion, not evidence and logic. It is very difficult to argue with someone's belief, and phrasing it in this way implicitly moves the discussion into the realm of the subjective, suggesting there is no one right answer to the question at hand – which might be acceptable in a newspaper opinion piece, but is not in an academic essay. It might be possible to argue this line about Donald Trump, but to do it in an academic context you have to mobilise sufficient evidence to support that position, and not simply revert to subjective assertion (albeit a very Trumpian move in itself).

Finally, the use of phrases like 'I think', 'I believe' or 'in my opinion' sounds tentative and even apologetic. As above, academic writing is a contribution to a body of knowledge on a subject, and that contribution should be based on what you can prove or justify, not what you think. Your voice is worth listening to, and you are capable of answering the question. While it is important to hedge and think about how certain you want to make your statement, you do not have to apologise for making the statement in the first place.

The first person is also not just about the use of 'I' – there is also 'we' to consider. What is the problem here?

- When considering colonial history, we have a tendency to have always already made the conquered/colonised victims and inferior and not consider them as truly having agency.

The question that you should always ask when you come across the word 'we' (and not just in an academic context) is – who is 'we'? Using the first-person plural implies a commonality between the author and the reader, and also a wider group. It universalises the point being made and makes it feel widely held. In this statement, it is likely, however, that any 'we' being invoked is a white, Western, colonising 'we', and one which therefore excludes a huge proportion of the world.

Of course, there are times when 'we' can be appropriate. If you are writing up research carried out by a group, then of course the referent of the 'we' (i.e. who is included) is very clear. Equally, in certain disciplines, 'we' (or even 'you' in some cases) is used to discuss certain things – for example, the workings of thought or as a convention when the text is asking for the reader's participation. I have done that repeatedly in this book, for example, saying things like 'in this chapter, we have looked at ...'. Again, the referent is clear – it is you the reader, whoever you may be, and me, the writer, and so there is no problem.

When you use 'I' in your writing, then, you should ask yourself whether it makes your argument (seem) subjective, and when you use 'we' you should ask yourself who are the other people that are included in your 'we', and whether you can truly make a claim to represent them. This is not an arbitrary or blanket rule in the way it is often presented, but rather a decision that is made based on the context you are writing for and the questions we discussed above.

What do you do if you don't want to use the first person though? Here, the passive voice can help us.

To passive or not to passive

The passive voice is another common feature in the teaching of academic writing, where students have it drummed into them that this is the best way to achieve appropriately objective and academic sounding prose. Consider the following:

- After adding 5 ml 0.5 mM NaCl I incubated the extract at room temperature for 2 hours.
- After adding 5 ml 0.5 mM NaCl the extract was incubated at room temperature for 2 hours.

The second is in the passive, and as you can see, the idea behind using this in academic writing is that it places the emphasis on the action and that which is acted upon rather than the actor. It removes the individual from consideration, which, so the argument goes, must make things more objective.

However, in many disciplines, and in many different types of academic writing, focusing on the individual actor may be the most appropriate thing to do. For example, in reflective writing, where a student is asked to apply theoretical frameworks to their own experience, or when taking an iterative approach to qualitative research, the use of the passive could make things more confusing while adding nothing.

As before, the key thing is to think about what you want to achieve by making a particular linguistic choice. As long as you can justify your choice, there is really no rule that you cannot break.

How do I know what is formal?

This is especially difficult for anyone studying in a different language, but academic writing can feel like a different language even for native speakers. While in some ways this is something that just requires time and exposure, and it is also worth noting that there is a longstanding trend away from strict formality both in academia and the wider world, there are some tips that can help.

- One-word verbs tend to be more formal than phrasal verbs in English. So look into → examine; look up to → admire; be down to → caused, getting near to → approaching, and so on.
- Avoid phrases that are actually metaphors, as these tend to be more informal – for example, see above where I described the passive voice as being something that students have 'drummed into them', and earlier I used

the phrase 'see the bigger picture'. These are so common in British English that native speakers tend not to realise that they are in fact metaphorical, idiomatic expressions that are not transparent to, for example, non-native speakers, and feel more informal due to their lack of precision.

- Get rid of adjectives and adverbs. Often, these are unnecessary in academic writing and as well as sounding informal, tend to make things less precise. Consider the 'diving beneath the waves ...' example we looked at above. Here, even if you just deleted the adjectives, it would sound more formal so – 'Diving beneath the waves in the waters of the equator you may be met with a view of a diverse marine city, inhabitants of which swim with little to no effort through their metropolis under the surface of the sea'.
  - If you know you have a tendency to use (or overuse) adjectives, make an effort to go through and remove them as part of your redrafting process.
- Learn from what you read. This is a good way to get a feel both for what is and isn't acceptable in your discipline in terms of formality, and is also one of the most important things you can do to improve your writing.

When reading, think like a writer

The most important resource you have when learning the language of argument in your discipline is the language of others in that discipline. That means the language that your lecturers use, but also, vitally, all of the reading you do as part of your study. Studying at university used to be called 'reading for a degree', and it remains a useful way of thinking about what you are doing as an undergraduate or postgraduate.

The most simple way that we all learn to speak different languages, whether those actually be foreign tongues, or just different dialects within our own, is through exposure and imitation. Again, this is something we all do automatically – you know how to talk to a policeman, or a teacher, or a waiter in your culture because you have learnt to do so by watching others and copying them.

The same process is at work when you are a student, and if you do it consciously you are likely to learn much more quickly. That means that whenever you read, you should be thinking like a writer – that is, thinking what you can learn from the text that you are looking at and what you can take and use yourself.

This will teach you a lot of different things. For example, you will learn a lot about what terms are and aren't used in your subject, what key terminology there is and what level of formality is common. You will also, just as importantly, learn about the conventions for structuring an argument, what sort of evidence is used to support what sort of point, and how questions are framed and answered (an example of reading like a writer is included in Chapter 9).

Engaging with this on a personal level will enable you to develop your own voice, as you will be able to think about what sorts of expression you think are particularly effective, what structures work well for different types of argument and so on. Imitation in this context is not a bad thing, it is the best way to learn. As Pablo Picasso reportedly put it, good artists copy, great artists steal.

Break any rule rather than say something outright barbarous

The other benefit of thinking about good academic writing in terms of responding to certain key questions and applying principles, is that it also allows you to break any rule you want, as long you are able to justify it in terms of those principles. To paraphrase George Orwell, it is better to break any rule than say something outright barbarous.

Consider the following example from David Graeber and David Wengrow's *The Dawn of Everything* (2021), where they are arguing about the history of the 'myth of the stupid savage', whereby so-called 'primitive' peoples were argued to be essentially lacking in the ability to rationally shape their own lives:

> This is not the place to document how a right-wing critique morphed into a left-wing critique. To some degree, one can probably just put it down to the laziness of scholars schooled in the history of French or English literature, faced with the prospect of having to seriously engage with what a seventeenth-century Mi'kmaq might have actually been thinking. To say Mi'kmaq thought is unimportant might be racist; to say it's unknowable because the sources were racist, however, does rather let one off the hook.
>
> (Graeber and Wengrow, 2021, pp71–2)

As we can see, from the point of view of many of the 'rules' above, this is not acceptable academic writing. It uses informal phrases ('let one off the hook'), for example, and it is emotively scathing about people holding a different argumentative position (who are lazy and/or racist). It is also, however, remarkably effective at making its point and the way in which it breaks the rule adds to the effectiveness, rather than detracting.

One way of describing this manipulation of style for effect is *rhetoric*. We will look at this a little more in the next chapter when we consider speaking, but for now, it is a key reminder that while all of the discussion is relevant to good academic writing in general, it is especially important when considering how to build effective arguments.

All disciplines are a set of questions about the world, and often the very complex and difficult issues that you are exploring are the result of, or born out of, much simpler, more everyday questions. The purpose of academic writing, or any academic communication, is to clearly communicate the answers to those questions. The key for you as a student is to understand what impact your choice of words has on your reader, and to be actively and consciously making decisions about what choices you will make to get your message across on any given occasion.

Summary

- The effectiveness of arguments in an academic context is not just about what they contain, but also about how they are communicated.
- Learning how to write good academic arguments is not about simply learning a list of rules that can be universally applied to all subjects, it is about making decisions in each individual context about:
    - Your purpose – what you want to say and why you want to say it.
    - Your audience – who you are writing for and how that informs both your purpose and your style.
    - The role you are playing – who you are pretending to be (or embodying), and what that tells you about the voice you should be writing and speaking in.
- Academic arguments are not about giving your opinion but rather about giving your answer, which is based on logic and evidence. It is important to remember that a good argument gives *your* answer, however, and doesn't just recycle or repeat the answers of others.
- A university student is an academic-in-training, and as such, a part of the academic community, and has to write accordingly. That means that you have to take responsibility for what you argue both in terms of its content and in terms of the linguistic choices that you make when expressing that argument.

FURTHER READING

Chapter 7: Getting to grips with rhetoric from Chatfield, Tom, *Critical Thinking* (Sage, London, 2018).

Thomas, Francis-Noel and Turner, Mark, *Clear and Simple as the Truth: Writing Classic Prose* (Princeton University Press, Princeton, NJ, 1994).

CHECKPOINT

Think about the subject you are studying or are going to study at university and consider the following questions:

- What do you know (or do you expect will be the case) about the sort of academic style that is used in your subject?
- What sort of genres are you going to have to write (essay, report, reflective, ... ?) and will you have to write differently for those different genres?
- What sort of evidence does your subject expect you to use? What sort of claims can you make on the basis of that evidence? What sort of claims can you *not* make?
- How do you feel about being part of an academic community of equals? Is this idea going to be challenging and uncomfortable for you, or easy? What does it mean to treat your lecturers as your equals?

NOTE

1. A good example of 'classic style' in use can be seen in the extract from *The Dawn of Everything* (Graeber and Wengrow, 2021) included at the beginning of Chapter 8.

8

Arguing in class

While a lot of focus is often placed – including in this book – on how to write good arguments at university, it is also vitally important to consider how to argue well verbally. Speaking is one of the key skills for a university student, and whether it is debating with your classmates, discussing ideas with a lecturer or giving a presentation, you need to be able to build and respond to arguments using the spoken as well as the written word.

The difference between speaking and writing has a key place in the history of thought in what is often called the 'Western Tradition' – that is the history of philosophy based on the work of Greek and Roman classical thinkers and developed in 'the West' (a geographically inaccurate term that is usually used to cover Europe, America, Australia and assorted ex-colonial outposts). It is perhaps tempting these days to think of speaking as the lesser of the two, but in fact, for much of intellectual history the opposite was the case, with speech privileged as being the place where arguments were truly 'present' and writing seen as cut off from the source of its meaning (the author).

While it is not useful for us to pursue this line of enquiry here, or to consider how structuralist and post-structuralist thought has changed the ways in which we think about what text, and indeed speech, can be said to mean, it is perhaps useful to consider the place of speech in the ongoing history of argumentation and knowledge.

David Graeber and David Wengrow (2021), for example, argue that human knowledge is fundamentally dialogic, by which they mean that there is a fundamental tendency for mankind to create knowledge in dialogue with an other – whether that 'other' is real or imaginary.

It is worth reading the passage where they discuss this in full:

> Philosophers tend to define human consciousness in terms of self-awareness; neuroscientists, on the other hand, tell us we spend the overwhelming majority of our time on autopilot, working out habitual forms of behaviour without any sort of conscious reflection. When we are capable of self-awareness, it's usually for very brief periods of time: the 'window of consciousness', during which we can hold a thought or work out a problem, tends to be open on average for roughly seven seconds. What neuroscientists (and it must be said, most contemporary philosophers) almost never notice, however, is that the great exception to this is when we're talking to someone else. In conversation, we can hold thoughts and reflect on problems sometimes for hours on end. This is of course why so often, even if we're trying to figure something out for ourselves we imagine arguing with or explaining it to someone else. Human thought is inherently dialogic. Ancient philosophers tended to be keenly aware of all this: that's why, whether they were in China, India or Greece, they tended to write their books in the form of dialogues. Humans were only fully self-conscious when arguing with one another, trying to sway each other's views, or working out a common problem ...
>
> (Graeber and Wengrow, 2021, pp93–4)

There are many interesting aspects to their argument here. Firstly, the assertion that humans spend an awful lot of time reacting to and interpreting the world automatically links to the discussions in Chapter 3 about the ways in which cognitive biases affect our ability to construct sound arguments. The idea that we can overcome this tendency and think more deeply and clearly when in dialogue with another is a fascinating one, and has clear implications not just for how we think about the construction of knowledge, but also the importance of speaking at university.

Secondly, Graeber and Wengrow (2021) are also suggesting an interesting model of the self here – by which I mean just a way of thinking about (a 'model' of) what it means to be a person, a 'me' or a 'you' (a 'self'). If someone is only fully self-conscious through dialogue with another, then there is a sense in which our 'self' is located not in any essential or discrete body, but is in fact part of a network that necessarily involves those others. There is something paradoxical about being only able to be fully aware of yourself when interacting with an other, but this is perhaps a useful model for thinking about how academic knowledge is not a static body of established objects ('facts'), but rather a conversation – a collection of shared interactions and not a fixed property of one person or group. This is a conversation in which you as a student participate, and the subject you study is therefore everything that is written and said about it, and not something separate which that writing and speech is talking about.

In the last chapter, I asked you to consider and construct your own imaginary reader so that you could make the best linguistic choices and build the most effective arguments when you are writing. This approach fits nicely with the notion that knowledge is constructed dialogically.

Of course, when we are actually speaking, we are usually speaking to a very real other person or people and not making them up. However, there is a sense in which even when engaging in 'real' conversation or debate, we are still engaging with an imaginary other, both because we cannot truly know what someone else is thinking, no matter how hard we try, and secondly because everyone in an academic context is occupying and projecting a role, which necessarily comes with certain functions in different situations. In other words, when we talk to people we only have access to the role they are playing and the language they are using, and we fill in the gaps to try and work out their argument and intentions, just as we do when we write not knowing who will ultimately be our reader. It is also worth reiterating here that in an academic context, we are constructing not just an imaginary other, but also an imaginary self – that is, we are taking on and inhabiting a particular role, and it is one that is as much constructed as the imaginary audience we are talking to.

What all of this means for us here is that speaking and writing are not opposites, with one to be privileged over another, but are instead all part of the same process of knowledge production and transmission, which we have argued throughout this book is the primary purpose of universities. The techniques, approaches and questions that we need to think about arguments are the same when speaking as those we have discussed in the rest of this book, then. Questions of validity and soundness, underpinning theoretical frameworks and clear, logical structuring all still apply.

What is different and what we therefore need to think about here is how the change of context (i.e. speaking rather than writing) changes the language that we need to use and the ways in which effective structure for spoken arguments are different to effective structures for writing. In this chapter, we will look at that in a little more detail, first in terms of general discussion and debate, and secondly in terms of giving presentations, or perhaps more generally, public speaking.

Speaking in the classroom

Discussions of speaking at university often focus on giving presentations, mostly as a reflection of the level of anxiety felt by many students about that prospect. However, presentations are in fact most likely to be only a tiny fraction of the spoken words you utter during your studies, and in comparison to all those millions of other words that you will speak, among the least important.

No matter what subject you study, much of the learning that you do, the ideas that you come up with and the knowledge that you gain will come through discussion with others. Whether it is through seminar debate, asking questions or simple conversation, speech is a key medium through which you grapple with the content of your discipline.

In order to think about how to most effectively build arguments in this medium, it is first necessary to think about what you are *not* doing. In academic conversation, you are not (or not very often) trying to 'win' – academic argument should not be gladiatorial. You aren't trying to beat the other person or people. At worst, you are persuading and at best you are openly considering ideas in a situation where lots of different perspectives and positions are valid. Think of it as *discussing with*, rather than *arguing against*.

One useful way of attempting to embody this approach is through the *Rogerian model*. This is discussed in detail in Chapter 2, but for our purposes here it is useful to remind ourselves that, in Rogerian argument, the most important thing is not to prove our own point, but rather to first 'listen with understanding' and attempt to understand not just another person's argument (that is, their point, or the content of what they say) but the worldview that underpins it (or in other words, *why* they are saying it). Key in this approach is that rather than just thinking about how to 'beat' the opponent, or how to find flaws in their reasoning, you are just as invested in understanding their point of view, and trying to fully occupy their position, as you are in advancing your own. This makes you much more likely to fully engage with their argument, and perhaps ironically, much more likely to see any potential problems with it (and therefore 'win').

It is easy to see how the approach discussed above, where we actively imagine and construct the others that we are speaking to, even if those 'others' are actual people that are right in front of us, is a useful one for the Rogerian model. But while it is easy to say that we will practise 'empathetic listening' with sympathetically imagined counterparts, it is much more difficult in the moment to abandon long-entrenched habits of debate, and let go of the emotional investment we have in our own points of view.

In order to actually do this, we need to change our actions, and one of the easiest things that we can do is to consciously change the language and the behaviours that we use in discussion with others. Before we look at some examples of how we can do this, think about the following question.

TASK 8.1

Who in your life do you enjoy (or have you enjoyed) discussing ideas with? This doesn't have to be academic ideas – it could be thoughts about football, or computer games, or politics, or books that you like, or the latest celebrity gossip. The person could be a relative, a teacher or a friend. I would simply like you to think about someone (or more than one person) that you have had good conversations with, and what it is that made or makes those conversations enjoyable. Of course, some of it will be about what you discuss – the content – but a lot of it will also be down to *how* you discuss it.

Ask yourself, what does that person do in conversation that makes it good to talk to them? What language and phrases do they use? How do they behave? How do they treat you? How do they react to your opinions?

Now ask yourself all the same questions about yourself. How do *you* behave in conversation with that person? Do you mirror their behaviour (that is, do the same things that they do or talk in the same way) or do you do it differently? Do you talk and behave in conversations with them in the same way that you do with other people? If not, why not?

Jot down a few of the things that you think are important. It's likely that what you value about conversations with this person will give you some clues as to how to be a good interlocutor yourself, including in an academic context. One important thing to remember in this, as in all contexts, is – you cannot change the other person's behaviour, but you can change your own.

Approaching academic discussion

How, then, can we approach academic discussion in ways that are collegiate, collaborative and constructive? The first important thing is to think of academic discussion in these terms, and to remember the most important rule – namely, you should always argue with the idea and not the person.

The second thing to remember is that we are as opaque to the person we are talking to as they are to us: neither party knows what the other is thinking, and so the way that you will come across is determined not by your intentions, but by the words that you use and the behaviour you exhibit.

The following are a few suggestions for approaches you can use to approach academic discussion in the right spirit. Remember that the qualities you must ascribe to the person or people you are speaking to are the same as the imaginary reader that you write for – that is, they are intelligent, capable and knowledgeable. They are interested in what you have to say and you are interested in what they have to say. This is a conversation between equals, and one carried out in a spirit of mutual respect and professionalism.

Everything here is underpinned by these fundamental principles.

Be an active listener

It is all too common to use the time when someone else is talking not to think about what they're saying but rather to prepare what you are going to say next. However, for academic discussion to be genuinely fruitful, you need to really *listen* and to try and take on board what they are arguing.

For example, while listening to someone, you can attempt to identify the assumptions/theoretical frameworks that underpin a person's argument (see Chapters 3 and 9). This will help to ensure that you engage with the position and not the person, which will tend to make things less contentious and more likely to lead to productive debate.

This approach is useful in various ways. An active attempt to contextualise someone's argument allows you to understand it more deeply, and makes it more likely that you'll be able to make links with ideas and knowledge that you already have. You also do not have to have detailed subject knowledge to do this as, similar to the trigger questions that we discussed for reading in Chapter 4, there are certain questions that you can ask in any academic situation. For example:

- Why has the other person taken this position? Is it motivated by a particular theoretical outlook?
- Is it based on a cultural perspective or other belief?
- Can I think of any other theoretical approaches that interpret things differently?
- Does the argument relate to any of the assigned reading on this module or elsewhere on my course?
- Can I make links between what is being discussed here and anything else that has come up on this module or course?
- Are there any counterarguments or alternative positions that I am already aware of?

By taking this approach, you can also make any potential challenge you have hypothetical and not personal. For example, saying, 'But I think a feminist would respond to that by saying that the argument hasn't taken into account ...', or 'this week's reading suggests that the opposite might be true' sounds a lot less confrontational than something like 'you clearly haven't thought of x', 'I disagree' or even 'you are wrong'.

Create a constructive space

As this demonstrates, being an active listener is not just about what is going on inside your head, it is also about what you say and do. Signalling to the other person that you are listening to them and want to hear what they have to say is vitally important.

You can do this in a number of ways:

- *Give credit to others* – Use phrases like 'that's interesting', or 'I'd not thought of it like that' – even when disagreeing.
- *Build on what others have said* – Your argument is more likely to be accepted if others feel that you have taken their points into account. Your argument is also likely to be stronger if you actually *have* done this. Many people are good at acting like they listen to others – it's much more difficult to actually *do* it. Simple phrases like, 'as x mentioned earlier', 'I agree with what y argued', or 'I thought z made an interesting point', demonstrate clearly to others that you have taken what they said on board, and also place whatever you want to say in the context of the ongoing discussion.
- *Turn taking* and making sure you *give space for others* to contribute – Think about whose voices are most often heard in a class, and whose aren't, and about what you can do to redress that balance. If someone else is dominating the discussion, don't just counter that by arguing back yourself – invite others to say what they think too.
- *Ask open questions* – This is a way of drawing people into the discussion, seeing what they think and allowing people to feel heard and included. Example phrases include 'Do you think that's convincing?' or 'Is that what other people took from the reading?', but the single most important one to remember is: 'What do you think?'. Good listeners ask questions and the best way to show someone that you're interested in their point of view is to ask them about it.
- *Give a right of reply* – If you add to, or disagree with what someone else has said, check with them afterwards if you've done them justice: 'Does that answer your question?', 'Is that what you meant?', 'Do you think that's a reasonable criticism?', 'Do you want to come back on that?', and so on.
- *Mirroring, summarising or asking for clarification* – When you respond to someone else, it can be helpful to summarise, or paraphrase what you think they have said. This is actually the exact thought experiment that Carl Rogers (2017) suggests as a way of approaching a debate (see Chapter 2), namely

instituting the rule: 'Each person can speak up for himself only after he has first restated the ideas and feelings of the previous speaker accurately, and to that speaker's satisfaction'. As well as taking the emotion out of the discussion, as Rogers says, it also allows you to check your understanding and to again root the debate in an engagement with ideas, and not in a conflict between individuals: 'So, I think you're saying *x*, is that right?', 'The argument you're making suggests *y*, and while I agree in part, it also leads to this problem …', 'Are we saying, then, that this is how we should think about x?', and so on.

- *Hedging and allowing uncertainty* – When discussing ideas, it is very tempting to present our own points in the most convincing way possible and to speak with a certainty that we may or may not actually feel. However, for academic conversation to truly be called a discussion or debate, we should be leaving open the possibility of different positions or of being wrong. Hedging – that is, using words like 'might', 'seem' or 'could' to present our ideas as possibilities and not certainties, and explicitly admitting your own uncertainty ('That's how it seems to me, at least', 'My interpretation is this, but I'd be really interested to know what others think') – are ways in which you can make the discursive space more inclusive and less combative.

All of this will make it much more likely that the discussion will be constructive and whether a consensus position is reached or not does not really matter. You might finish any conversation you enter holding a different view than you did when you began it – and that is the point. Through engaging with others, we hold our concentration on an idea, we open up chains of association that might otherwise lie dormant and we allow links and connections to be made.

Discussion and debate is a creative act that is necessarily collaborative and thinking carefully about the language that you use in, and the approach that you take to, discussion will allow you to both be someone that others like talking to and a student that gets the most out of their studies.

CHECKPOINT

Before moving on to the next section, take a look at the suggestions above and think about which of these are already features of the way that you discuss ideas with others and which you have never tried or need to work on.

Identify *at least two* that you would like to adopt or improve on and the next time you are debating or discussing ideas with someone, see if you can put them into practice.

Giving good presentations

Presentations make up a relatively small amount of most undergraduate degrees but cause a disproportionate amount of anxiety. In a moment, we will look at ways in which you can address that anxiety, but first let us consider some general rules for giving good presentations as a student.

TASK 8.2

Before we start, take a couple of minutes to think about everything you know about what makes a good or a bad presentation. Also think about how an academic presentation might be different from other types of public speaking. Make a note of all the ideas you have, from the most simple (e.g. don't swear) upwards.

There are many things that you could have thought about in response to this question. Here are a few examples:

- Engage your audience:
  - ask them questions (rhetorical and otherwise)
  - make eye contact
  - speak at a reasonable volume
  - don't talk too quickly
- Use good visual aids:
  - don't include too much text on slides
  - don't use distracting fonts, colours or backgrounds
  - only use images if they have a very clear purpose
  - don't have too many slides
  - remember that visual aids are there to support your presentation and not *be* your presentation; the most important thing is what you are saying – always
- Be academic:
  - use a formal and professional tone
  - use academic terminology where necessary but keep your language as clear and simple as possible; spoken and written English are in many ways different languages that follow different rules – remember that you do not speak how you write

    - you still need to back up what you say with evidence; presentations should be referenced in the same way as academic writing

- Be prepared:
    - know what you want to say, and what you want your audience to take away – we will come back to this shortly
    - make sure you are going to stick to the time limit given – you don't want to run out of time before you've finished making your point
    - know where and how you are going to have to present, make sure the technology works and that you know how to use it, and know where the audience are going to be and where you are going to be – again, we will come back to this in a moment

These are all good pieces of advice on how to give a good academic presentation. However, what we are most interested in here is not general tips for giving a presentation (although these are important), but rather how to deliver *a good argument* in a spoken form.

Framing the list above in this context is important, as this is not just an arbitrary checklist that you are being judged against as a student, although it can often feel that way. Instead, it is important to stop and ask ourselves – why? Why have I learnt these rules about good and bad presentations? Where have they come from?

The answer is that all of these 'rules' come from answers to the question: how can I best communicate a complex academic argument while speaking? We speak at a reasonable pace as academic arguments are complicated and we need to ensure that our audience can follow what we are saying; we provide references because in an academic context we have to support everything we say with evidence; and we use appropriate academic terminology because we can assume an intelligent, informed audience, and so on.

As discussed in Chapter 7 with regard to writing, these are not actually rules at all. If you can justify why breaking them would be useful in order to better communicate your argument, then break them.

Structuring a good presentation

While there are key similarities and overlaps, delivering a good argument verbally is different from delivering a good argument in writing in some important ways. Delivering your argument as a presentation rather than as a piece of writing requires, most importantly, that you think about your audience. It is also important to think about the role that you are playing – but we will come back to that in a moment when we think about nerves.

Our audience, as we have already established, is the same as the imaginary readership for our academic writing. They are an intelligent, capable, interested and informed group of your peers – that is, your equals. You do not have to sell the topic to them or make it seem more appealing than it is. But no matter how educated or intelligent they are, it is a very different proposition to follow a complex argument as a listener than it is as a reader, and you need to give them as much help as possible.

This is why the structure of your presentation, both overall and within each section, is so important. Having a very clear sense of your overall argument is vital, not just in terms of planning but in terms of actually being able to express that clearly and concisely.

One way to do this is to try and some up your entire argument in one or two sentences. This is sometimes referred to as an 'elevator pitch', or in other words, what you would have to do if you had to deliver your whole presentation in the time it took to complete a single journey in a lift. This is a useful exercise when writing but is even more useful when presenting. You should be able to tell your audience exactly what you want them to take away from your presentation, and if you are confused, they will be too.

Because of this, it is essential to have a clear, simple structure, if not in overall terms, than at least in terms of how you lead your audience from point to point. They cannot pause you, or rewind you, or stop to consider what you've said. They need you to lead them by the hand.

The structure itself might be very similar to the one that you use in a piece of writing then – that is, the points you make might come in the same order. What is likely to be different, however, is the amount of *repetition* and *signposting* that you include. Repetition is fairly obvious – by repeating important points, we ensure that the audience notes and understands them. *Signposting* is perhaps a less well-known term, but it is again a very simple idea – it simply means telling your audience what you are doing at each point and what is going to happen next.

TASK 8.3

Take a look at the following example, taken from the beginning of a lecture called 'The Melting Pot', which I delivered in 2018 as part of a module called 'British Society and Culture', which was designed for international students spending a year or term studying in the UK.

See what strikes you about the language that is used – both in terms of where you think I am 'signposting', but also where there are any examples of other ideas that we've discussed in this chapter so far:

Today I want us to start to elaborate on, and pick apart, some of the broad themes that we looked at last week. So rather than look at one particular cultural object or period in detail, I would like to take a very broad approach and start to think about what British culture is, what it is presented to be, and why that is the case – historically and now. I would also like to suggest that the version of Britain that is often presented is one that is both quite accurate, and at the same time, should not be given any great credence – both because it does not exist, and because, as we shall see, trying to match actual, concrete examples of a culture to the abstract, theoretical concepts that are supposed to underpin it is exceptionally difficult. Basically, what I want to do is look at some of the foundational narratives that Britain likes to tell the world, and itself, about itself (and to give a context for all of the things that we're going to discuss for the rest of the term).

As part of this, I would also like to have a look at how much of British culture is not 'British' at all. Are you familiar with this phrase 'melting pot'? It's actually a bit of a controversial one, and one that I used a bit glibly when I was designing this week's material, but I hope you'll be able to see why … It contrasts with the usual word that is used to describe Britain these days, which is 'multicultural' (more later – and to be discussed in the seminar). I'd like us to think about what is at stake in these metaphors – and others, like 'motherland' or 'fatherland', like 'the American dream' or … (more examples – see if they have any?)

First, I want to take a look at what stereotypical 'Britishness' is here …

Here are some things that you might have noticed:

- The signposting is very explicit ('I would like to ...', 'First, I want to ...' etc.). I am telling the students what is coming up in the lecture, and showing how what I am talking about in this lecture links to the previous week.
- The main argument of the lecture is laid out here, even if not explicitly – 'the version of Britain that is often presented is one that is both quite accurate, and at the same time, should not be given any great credence' – and key elements are repeated: we are going to be taking a broad overview, not deal with specific examples; we are going to look at metaphors and narratives concerning Britishness, etc.
- The language is quite informal ('it's actually a bit of a controversial one'), but it is not dumbed down (e.g. 'trying to match actual, concrete examples of a culture to the abstract, theoretical concepts that are supposed to underpin it is exceptionally difficult').
- Note, also, that this is a complete script, but that I have tried to write that script in a way that I will be comfortable saying out loud. I also include spaces for improvisation (e.g. '(more examples ...?)'), and notes to myself ('and to give a context ...', 'more later'). I do not read from the script when I deliver a lecture like this, but having written it out in full helps me to deliver it well, and having the full script is a comfort blanket, as I know that if I forget what to say, I have it right there in front of me.
- I use hedging constantly – 'I would like to *start* to', 'I would also like to *suggest*' – and I am not, therefore, claiming that I will completely answer the questions I am setting out. This explicitly frames the lecture as the start of a discussion with the students, and while an undergraduate presentation is different from a lecture, the same principle applies. Framing your argument in this way also makes it harder to argue against, as you are not including any definitive statements that are vulnerable to the black swan response (see Chapter 2).

There are other phrases that you can use to signpost as you move from point to point or between sections. As at the end of this short extract, you can simply number your points: 'First, I want to ...'. You can also signal when you have finished one part and are going to move on: 'Having looked at x, I would now like to talk about y', and so on.

At all points, giving short summaries, whether of what you are about to say, or what you are going to say, is also a really good way of ensuring that your audience stays with you and understands your points ('In other words, I would like to ...'). It is also important to lay bare the logic of how your argument fits together – that is, why your propositions lead to your conclusions and how everything joins up.

Take a look at another short extract from a lecture on the same module, this time called 'The Shakespeare Myth', which is about the role that playwright and poet William Shakespeare plays in British culture.

Firstly, the introduction:

> I want to do two things today – I do want to tell you something about Shakespeare, as he is a figure that looms so large in British culture, and it seems almost perverse to ignore him when you are doing a course like this; but more importantly I want to look at why Shakespeare has that status, why he is such an important figure, and more broadly why every culture has these icons.

As you can see, this is much less wordy than the previous introduction I showed you, but it serves many similar functions, including sign-posting what is coming up and establishing the key question that will be looked at – namely, why is Shakespeare such an important figure in British culture?

Next, take a look at this extract from the conclusion:

> So it's not really about what Shakespeare wrote – it's about what he stands for. Most people, beyond what they were forced to study at school, will never read him, or see any of his plays anyway. And if they do, they are consuming a commodity, they are acquiring cultural capital as much as they are connecting with any 'authentic' Shakespeare.
>
> Yes, Shakespeare was a great writer – but he is not the 'natural genius', the storyteller with the uncanny ability to see inside the human soul, the timeless artist who has found the plots (and the words) that explain human life. These are myths that become nonsensical under even the faintest scrutiny. How relatable is being asked to revenge your murdered father by his ghost? Or your teenage romance ending in a double suicide? Or your amateur dramatic production ending in you being kidnapped for the entertainment of the king and queen of the fairies? If everyone is raised being told that your plots are archetypal, then of course they will seem timeless ... And that's ignoring the fact that Shakespeare stole most of his plots anyway!
>
> Equally, if his use of language really was 'timeless', or capable of speaking directly to the soul, then why would we have to teach schoolchildren how to understand it, or publish editions with extensive notes to help readers translate it?
>
> I'm being facetious – but the point is that by the end of this course I do not want to see any of you uncritically buying into the myth of *any* cultural

icon, whether British or otherwise. With any cultural icon – or just figure that comes to occupy an important role in a culture – there is the reality, and then the story, and the two are very hard to disentangle. Most often what is being discussed is the story, the myth – what that figure *means*, *why* they are important, and what that says about the culture. That is often a long way from the 'truth' of what can be said about the individual in question, and at the very least will be highly selective. I would ask you to consider which figures occupy privileged positions in your culture, and why, and to consider whether the picture your culture paints of them really matches the reality, or whether that picture in fact says more about the culture than it does about the icon.

As you can see, the conclusion summarises some key points, before coming to a final point that very explicitly tells the audience what the lecture is trying to say.

You might also note how many rhetorical flourishes there are for effect here. The language is much stronger and emotive ('nonsensical under even the faintest scrutiny'); there are frequent rhetorical questions, exaggerations to make things seem ridiculous ('your amateur dramatic production ...') and repetitions of the same ideas.

Equally, however, there are still enough caveats in here to hedge the argument – 'Shakespeare was a great writer'; '*most often* what is being discussed ...'; '*at the very least* will be highly selective', and complex ideas and terms are couched in simple, informal language, or paraphrased when necessary.

The other key thing here is that the argument of the lecture is framed throughout in terms of the audience – why it matters to them, how it fits into the purpose of the course, and ways in which they can apply that argument to other situations.

Your purpose when giving an undergraduate presentation will be different from those of a lecturer, but the same principle will apply – considering your argument and how you express it through the lens of the audience is the best way to get your message across clearly and effectively.

Nerves

As already discussed, presentations are a very common cause of anxiety among students. There are many techniques out there that you can try, from embracing your nerves, to breathing exercises, or even 'power posing'. This last involves spreading your legs and lifting your arms above your head, or otherwise adopting a powerful stance, in the hope that assuming confident and powerful body language will actually make you powerful and assertive. There are TED talks on this, and books have been published on the subject.

It is, of course, nonsense. However, there is a germ of truth in it, which we can take and combine with the ideas already discussed here to think about how you can address your anxiety – or appear to, at least.

At various points in this chapter, we have considered who our audience (imaginary or otherwise) is in an academic context. This is the point to consider the flip side of that question, namely who *you* are, and what a good academic presenter looks like and can be said to be.

The answer is the same as it was in Chapter 7, that you are an intelligent and capable individual, who knows about their subject and whose voice deserves to be heard – or at least that you must act as if you are all of these things, even if you do not believe you are them.

Acting is the key word here. And while it might sound like underwhelming advice to simply *pretend* that you are confident, it is perhaps more useful than it sounds and has the advantage of actually being possible. It is also something that you do every day. Every situation that you walk into, you (consciously or not) evaluate who you are talking to, what role you are playing and what you want to achieve in that situation. You make linguistic choices accordingly – you talk differently to your mother than you do to a policeman or your friends – and you also adapt your behaviour. If you are trying to convince someone to give you a job you behave differently than if you are trying to convince them you are a good romantic prospect, or even that you are a dedicated and intelligent student that is about to give a very interesting presentation.

You may not be able to 'fake it until you make it', as power posing or other self-help approaches might suggest, but you can at least consider how to fake it convincingly. You play different roles every day, without any difficulty, and while it is more difficult with activities such as presentations that make us nervous, there are things that you can do to address your anxiety.

Firstly, think about what elements of giving a presentation you can control. Think about what a confident and assured presentation and presenter looks like, in terms of how they dress, their body language and facial expression. Think about how nerves manifest themselves in you – if you know that your hands shake (like me), make sure, for example, that you are not holding papers at the beginning of your talk. If you know that you have a tendency to talk too quickly, have a prepared introduction that you can deliver slowly and calmly to allow the adrenaline to wear off. Make sure that what you've prepared to say is something that you are comfortable saying, and that you haven't set yourself up to fail by including difficult words or sentences.

In all aspects, minimise the chances that you will make a mistake. Be prepared, make sure you have practised, and make sure you know exactly what you want your audience to take away from your presentation. This last point is key as feeling secure in what you want to communicate and having a good sense of your overall argument is the best way to calm your nerves. You may not be sure of your expertise, but you can make sure you have thought about why you are giving your talk – is it an assessed presentation with a clear mark scheme that you have planned for, or a seminar paper designed to discuss a particular question or piece of reading? Is it a proposal for research or a report on research? Is it a general introduction designed to start a discussion? Each does something different, and having that purpose clear in your mind makes you much more likely to fulfil it. Having a clear, one-sentence summary of your argument that you have memorised is also great fallback, and even if you get a bit muddled in the middle, it gives you an anchor point that you can return to at the end.

Overall, remember the role that you are playing and the qualities that you are supposed to be demonstrating. What does a professional, intelligent, reasonable, objective person talk like? How do they present their arguments? What will make you feel comfortable and like you are projecting these qualities?

Remember also, to map the qualities that we discussed in our reader onto your audience. Underlying the fear of public speaking is often the worry of being embarrassed or laughed at. But remember, this is an audience of equals, they are interested in what you have to say and they are not hostile. They want you to do well and they want to understand your argument. Help them to do so.

Of course, none of this means that you will actually stop feeling nervous. It might, however, give you the ability to act as if you are not.

The following checklists might help:

Checklist – the day before:

- Do I know exactly what I want to say? Can I sum up my argument in one or two sentences?
- Have I practised and am I within the time limit?
- Have I read my presentation out loud at least once (with or without an audience) to check that I am comfortable with all the language included?
- Do I know my purpose?
- Have I checked that I am following any instructions and/or meeting the marking criteria?

Checklist – on the day:

- Do I know where the presentation will be given and how the room is laid out?
- Have I checked the technology?
- Do I have a print-out of any notes or script and a working copy of any visual aids?
- Do I have a bottle of water with me if I need one?
- Have I thought about any questions I might be asked?

If you can answer 'yes' to all of the above, then the chances are that even if you still feel nervous, your anxiety will not stop you from delivering an excellent presentation.

Summary

- Speech is not only a functional skill, it is also a key part of the academic process by which knowledge is created and communicated.
- Speaking is a vital skill at university, whether in class, in discussion with other students or when delivering presentations.
- In academic debate, you are always arguing with the idea and not the person. You should strive to be an active listener and consciously seek to create a collaborative space for constructive argument.
- In spoken arguments, it is even more important to be absolutely clear about your overall position and how that position is backed up. Signposting and repetition are important ways in which you can ensure your argument is clearly structured.
- When speaking, we need to consider both the audience we are speaking to, and the role we are inhabiting as speaker. Both are always, to an extent, imaginary, and thinking about the ways in which we construct both, through interpreting what we see in others and what we project ourselves, will help us to make our arguments as effectively and clearly as possible.
- If you consider your purpose, make sure you are prepared and have thought about what elements of your presentation you can control – you are much more likely to be able to control any nerves and deliver a good talk and a clear argument.

FURTHER READING

Oral Arguments from Weston, Anthony, *A Rulebook for Arguments*, 4th edn (Hackett, Indianapolis, 2009).

Van Emden, Joan and Becker, Lucinda, *Presentation Skills for Students*, 3rd edn (Bloomsbury, London, 2016).

9

Making the argument your own

University mark schemes, whether at undergraduate or postgraduate level, often talk about 'originality' or 'criticality'. To get a 'good degree' (that is, a 2:1 or above), it is not enough just to show 'knowledge of the material, arguments and sources', you also have to 'demonstrate powers of critical analysis and originality of thought', or 'independence of thought' and 'critical acumen'.[1]

All of this might sound intimidating, but it need not be, and how to satisfy this criterion is key in understanding both how to build good arguments and how to do well at university. Despite how it is often phrased, being asked to demonstrate 'originality' is not about coming up with something entirely new. The criterion for awarding a PhD is that a thesis makes a 'new and significant contribution to knowledge', but at undergraduate or even Masters level, you are not expected to go anywhere near that far.

As well as the question of how you can possibly come up with something original when so much has already been argued, there is also always the nagging question – how would you know? How do you know if something has been said before or not? It is not possible to read everything that has ever been said on a topic, even for the most dedicated of students (or academics). And the answer is – you can't. You can never know for certain if something you argue has been argued before, and nor are you expected to.

So, we are not expected to say something completely new, and we could never be truly sure if we have, even if we did. What then, are students expected to do in order to show originality and make an argument their own? In this chapter, we will explore three ways in which you can do this: firstly, looking at how to make sure that the evidence is there to support, and not *be*, your argument; secondly, looking at the ways in which we can problematise the questions we are answering, whether these be questions set by lecturers or research questions we have developed ourselves; and lastly, we will explore the creative potential of different approaches, and how taking them will affect your arguments.

Don't sweat the small stuff

The first and most important point to recognise here is that originality does not have to mean coming up with ground-breaking or earth-shattering ideas that completely transform the field. As we have established repeatedly throughout this book, much of what you will be expected to do as a student will involve engaging with the work of others and understanding, deconstructing and reconstructing their arguments in order to produce your own.

Indeed, the ability to take a large body of thought and provide a clear and concise account of it, and the arguments it contains, is in itself one of the key skills that university equips students with, whatever their subject of study, and one of most important features distinguishing the best students.

In this context, *doing anything that is not simply describing or repeating what is said elsewhere* involves an element of originality. Making comparisons, noting similarities and trends, identifying gaps (i.e. questions that have not been asked or areas not explored) – all of these are critical acts. Making links between arguments made in separate texts or playing one argument off against another likewise involves independent thought, even if what is produced is largely reporting on the ideas of others.

Doing all of the above enables you as a student to add to what others have said, and even if you feel that the comments you are making seem relatively small, this still constitutes the originality that your lecturers are looking for. What you are in fact doing here is making an argument, whether that be about what the sources you are drawing on are saying, what the difference between them means, or what they have failed to say. That argument is *yours* and is therefore original.

TASK 9.1

Consider the following extract from an article by Ken Hyland (2018) on the teaching of English for Academic Purposes (EAP) at universities around the world. EAP is the name of both the teaching of English to speakers of other languages at university, and the name for the discipline that studies that area of teaching and the related pedagogy. It is perhaps worth noting here that some practitioners in this field would regard this book as an example of EAP while others very much would not.

As you read, see if you can identify where Hyland is describing what others say and where he is adding his own critical and independent insight. Where do you think he is only demonstrating knowledge of what others say and where is he adding something original or new?

> The first perspective concerns … [EAP's] role in the unfettered spread of English to a dominant place in academic communication. This spread can be regarded either as essentially benign, a neutral lingua franca efficiently facilitating the free exchange of knowledge, or as a Tyrannosaurus Rex, 'a powerful carnivore gobbling up the other denizens of the academic linguistic grazing grounds' (Swales 1997: 374). For some it is the latter, a Trojan horse of Anglophone values and economic interests perpetuating reliance, removing choice and eroding linguistic diversity. But while EAP teachers and students 'now collectively occupy global, interconnected spaces' (Morgan & Ramanathan 2005: 152), they have done little to question or arrest this growth.
>
> While historical circumstances, largely the legacy of US and British colonialism and the expansion of a single market across the world, are responsible for this spread, critics argue that EAP has been complicit in it. Not only has EAP failed to address the global expansion of English, fuelling the loss of specialized registers in the interests of big business (e.g. Pennycook 1997; Benesch 2001), but profits from it by providing courses, curricula and materials to universities and students worldwide (Phillipson 1992; Canagarajah 1999). Indeed, some critics have directly implicated English language teaching in US-led imperialism. Kumaravadivelu (2006), for example, has argued that it stands behind capitalist expansion while Edge (2004: 718) has claimed that English Language

(*Continued*)

(*Continued*)

> Teaching (and EAP by extension) served to create conditions for the invasion of Iraq by facilitating 'the policies the tanks were sent in to impose'.
>
> Less dramatically, EAP helps fuel the frenzy of academic publishing in English. It is estimated, for example, that six million scholars in 17,000 universities across the world produce over 1.5 million peer reviewed articles each year in English (Bjork et al. 2009). Once again, although EAP is not responsible for this growth, it contributes to the conditions which make it possible. Almost all Czech, Hungarian and South Korean journals indexed by the Web of Science/Science Citation Index are in English for example, while France, Austria, Germany and Spain publish 90% of their journals in English (Bordens & Gomez 2004). Italian has declined significantly in academic publications (Giannoni 2008) and Swedish has virtually disappeared altogether (Swales 1997) ... By providing curricula and teachers to support this spread, EAP is held to be part of the problem.
>
> (Hyland, 2018, pp385–6)

Although, on the surface, it might appear that all Hyland (2018) is doing here is summarising or surveying the ideas of others – and he certainly includes a lot of evidence and references – there are actually surprisingly few points in this extract where he is not also demonstrating critical engagement. There is a clear sense throughout of Hyland's voice, and while he is drawing on sources throughout, he always does so in order to back up an argument and to make a clear point, rather than to simply repeat what they have said.

Before we move on to analyse the extract, I would like you to consider one more question, namely – what do you think Hyland's (2018) position is on EAP, now that you've read this extract? Does he think the criticisms discussed here are justified or not? What in the text would you use to support your answer?

Hyland starts this extract by summarising one trend ('The first perspective concerns ...') in what has been written about EAP over the last 20 years. To do this, he first sketches the opposition between those who regard the 'spread' of English as the dominant language in global academia as a positive ('this spread can be regarded either as ... benign'), and those who see it as a negative ('or as a Tyrannosaurus Rex'). He does not explicitly say which side of the argument he supports, but the lack of qualifying or **hedging** language in the final clause of that first paragraph (that is, Hyland (2018) states that EAP practitioners 'have done little to question or arrest this growth' as fact, he does not say that this is 'argued' or 'claimed' by someone else) clearly shows that he considers EAP to be a part of that spread. The slightly florid metaphor and choice of verbs elsewhere ('a Trojan Horse of Anglophone interests ... perpetuating ... eroding ...') does, however, point to some undercutting of the more extreme versions of this line of attack.

Hedging – Language used to qualify how strongly you want to argue or state something. Think about the difference between 'this suggests that' and 'this proves that' for example.

The second paragraph builds on the first, outlining critiques of how EAP has been 'complicit' in this spread of English. Again, there are clear signals of where Hyland is reporting on similarities and trends in the various sources he is drawing on ('critics argue that') and the lack of reporting verbs or hedging language in some sentences suggest places where he agrees ('Not only has EAP failed ...'). In others, distancing language ('indeed some critics ...') and reporting verbs suggest disagreement, with the 'has claimed that', for example, signalling a greater degree of disagreement than the 'has argued that', which is more neutral.

The final paragraph is the one place that Hyland (2018) could be seen to be offering no critical insight of his own. Here, he is mainly reporting on data through a summary of certain key statistics in academic publishing. However, there is still a very clear purpose, both in the choice of language and the choice of statistics. Firstly, the 'less dramatically' of the opening tells us that Hyland thinks that both the previous two points were a bit over the top and that what is to follow is more reasonable. Secondly, it is noticeable that Hyland focuses not on the countries that would perhaps be expected to be most affected by the 'legacy of US and British colonialism', which he has already said is responsible for the spread of English as lingua franca, that is the colonised countries themselves, but rather other European countries that would normally be considered part of the 'Western' dominant. Lastly, the final 'by providing ... EAP is *held* to' leaves us with a note of hedging through which Hyland suggests he does not consider this argument to be entirely fair.

Having read all this, we would perhaps conclude that Hyland agrees with many of these critiques of the role of EAP in the spread of English as an academically dominant language, but that he regards some of those critiques as being too extreme. In considering his overall position, however, we would also need to remind ourselves that the title of the whole article from which this extract is taken is 'In defence of EAP ...'.

Overall, the summary of Hyland's (2018) position that we have just come to would be a fair reflection of the rest of his article, where he goes on to say that 'it would be foolish to deny the detrimental impact the spread of English has had on other languages', and that 'EAP has helped strengthen this expansion' and 'been complacent in these developments' (p388). He also goes on to argue, however, that responsibility for this cannot be placed at the feet of EAP teachers themselves, as 'accusing marginalized EAP teachers of complicity in the consequences of the global spread of English might be likened to blaming coal miners for air pollution' (p389), meaning that the nuanced position we deduced from the extract alone is an accurate reflection of the argument behind the whole piece.

While these subsequent statements that Hyland adds are where he explicitly advances his own position, it is also very clear both how his argument is built on others' ideas, and striking how much of that argument we were able to identify just from his seemingly neutral discussion of those ideas. This demonstrates precisely the sort of engagement that university mark schemes describe when they talk of 'originality of thought' or 'critical acumen', and shows that it does not need to involve dramatic or expansive statements. Even just surveying other people's ideas, in other words, if it is done in a suitably critical way, constitutes presenting an original argument.

Problematising the question

The second way in which you can make an argument your own is to problematise the question. Academic arguments are always – explicitly or not – produced in response to a question, but those questions are themselves the result of other arguments, decisions and contexts. By considering these, we are able to think about how the way the question has been posed shapes the possible answers to it, and to consider the other possibilities that might result from changing the terms of that question.

At first glance, it may seem very obvious to say that the terms of a question shape the possible answers to it, but it is more complex than it first appears. Often, we think of setting the terms of a question as simply narrowing the focus to a manageable section of a topic, and while this is in itself more problematic than it seems, there are unseen assumptions within the way a question is posed that are more fundamental than this. Consider R.D. Laing's famous book *The Divided Self* (1960), for example, where he argues that how psychiatry understands madness is fundamentally flawed because it studies individuals as organisms rather than individuals, and 'man ... seen as an organism or man seen as a person discloses different aspects of the human reality to the investigator' (p22). He insists that 'both are quite possible methodologically', but his whole approach is based on showing what the latter approach reveals that the former does not. Most important for Laing is that considering man as a person necessitates considering individuals through their social context, rather than as individuals cut off from that context. He calls this choice of approach an 'intentional act', which anyone seeking to answer a question chooses 'within the overall context of what [they] are "after"' (p22). Changing the initial intentional act thus changes everything.

To think about this, let's work through an example. At the very beginning of this book, I asked you to consider the question 'why are you here?' and in particular, why you thought going to university was a good idea. In order to think about that question, we generated a lot of other questions that were related to it, one of which was 'what is a university degree worth?'.

Here, I would like us to consider that question in a little more detail. In order to do that, the first thing we need to do is to consider how a degree's worth is currently calculated in the UK. That in itself is a first engagement with the terms of the question (or intentional act), as we are implicitly narrowing its scope by considering it only in relation to one geographical area or context. It is worth noting that the way in which we problematise this question would be different if we were not to make this choice.

In the UK, an undergraduate degree for 'home' students currently costs up to £9,250 per year in fees. Very few courses cost less than that, and fees for international students are often substantially higher. That then is what a degree can be said to *cost*, but is that what it is *worth*? The two, surely, are slightly different questions.

The answer to this question is usually judged on the economic difference having a degree makes for the individual, with the Department for Education reporting in 2019 that graduates earned £10,000 per year more than non-graduates on average after graduating, and could expect to earn 20 per cent more over the course of their life-time (2020).

However, both of these figures represent an average, and clearly some degrees result in much higher earnings than others. The Office for

Students (OfS), which is responsible for the higher education sector in the UK, refers to these as 'high-value' and 'low-earning' courses, and is very alive to the question of 'value for money'. Those that are deemed 'low earning' – i.e. that produce graduates on lower salaries or not in professional roles – are currently under threat.

Here, then, we have an initial argument about what a degree can be said to be worth – it is about the economic benefit that it confers on an individual. Answering this question in the current UK context, you would have to take this position into account, as it is the one that has driven much government policy over the last decade and a half, and continues to shape the decisions of individual universities, and indeed students.

How robust is this way of defining the worth of a degree, however? Even if we stay only in the realm of the economic, we can see that there are various problems here. Firstly, a degree might allow someone to pursue a career (including those mentioned above) that are worth less in terms of salary paid, but worth more in terms of quality of life – it might be more fun to be a musician than a marketing executive, for example, even if the pay tends to be worse. Secondly, the economic value of a degree is not just to an individual, but to the wider society. For example, 'low-earning degrees' include teaching, nursing and social care, along with the arts. Do we want a society with less teachers, nurses, carers and artists? These jobs have a much higher social value than, for example, hedge fund managers, yet would be ranked as worth much less under the graduate-outcome focused model.

Privileging the economic value to the individual in this way creates, as Hooley and Mellors-Bourne (2020) note, 'what economists call a "perverse incentive"'. That is, 'in an attempt to incentivise one kind of behaviour (delivering quality higher education), we end up incentivising a quite different, and potentially socially destructive, behaviour (reducing the number of students on socially valuable courses)'.

This approach has also led to grade inflation, with the number of first-class degrees awarded climbing rapidly, in what the OfS themselves describe as an 'unexplained' way. The most likely explanation is that students as consumers demand a return on their investment, and when a degree is a product, the need to deliver a 'high-quality' commodity (i.e. a good degree result) places pressure on academic standards. This undermines another arguable value of university degrees – that is, providing a means to determine who are the brightest and best in different fields. Thus quantifying a degree in terms of earning power ironically makes that degree 'worth' less in terms of what it tells you about an individual's ability.

Responding to this argument, it is also clearly necessary to take issue with the idea that 'worth' is solely an economic question. In Chapter 1, I argued that a common way of looking at the value of university was on a binary scale with 'money' on one end, and 'knowledge' at the other. That opposition is just one way that we could call into question the notion that the value of a degree is based solely on economic output, and you might also think about many of the other factors raised in Chapter 1, including the social value of university study or the experiences it offers an individual.

I will not rehearse those arguments here as none of this is 'problematising the question', but rather simply exploring the question fully, accepting the terms on which it is based and proceeding accordingly. That is a valid approach, but it is also fruitful to take a step beyond that and to consider those terms themselves.

It is most common, as we have seen here, for responses to this question to argue whether or not we should determine the worth of a university degree based on the benefit it confers on graduates (i.e. the economic argument), or whether we should make our judgement using a different standard (the knowledge produced, the social benefit, the cultural capital, etc.). However, no matter how much we prefer one standard over another, we have to recognise that *all* of them are in some cases valid. Sometimes, measuring the salary benefit to an individual *is* the appropriate standard, and is in fact the reason why many students choose to go to university or to study a particular subject. The same is true of all the other benchmarks. Does this simply mean that this is an unwinnable argument (always a dubious case to assert) or is something else going on?

The problem is that in our discussion so far, we have interrogated the term 'worth', but we have not considered the term 'degree', assuming it, most probably, to be straightforward. However, even the most superficial consideration of the word tells us that this is not the case, and even some of the points raised here show us that the term 'degree' covers a wide range of different courses with wildly different purposes, methods of study and possible outcomes. Think about how you study a degree in medicine and what you expect to get out of it compared to a degree in pure mathematics or a degree in theatre production, for example.

Considered like this, it becomes fairly obvious that it is not possible to arrive at *one* scale that can measure the worth of all of these different degrees. That means that the question we are discussing is not one that it is possible to answer because it is based on a fundamental misconception.

This question has arisen in the first place because a particular way of thinking about the world (a neoliberal, capitalist model) has become dominant in wider society and is thus often seen as the only useful way of examining the world – including higher education. If we question that assumption – or choose a different intentional act – and decide instead that university study is not a commodity that can be neatly defined and that different degrees (even in the same subject) are not necessarily commensurate objects – by which I mean you cannot measure them against each other – then we produce a much more interesting position. Under its own terms, the question is impossible, and rather than engage in an argument on those terms that is ultimately empty and/or unresolvable, we should contest the way in which the topic is being considered in the first place. On a practical level, this would mean, for example, challenging government policy that sought to measure all degrees against a single standard and arguing instead for a different model – just as Laing did with psychiatry.

Putting it into practice

The approach that allowed us to problematise the question 'what is a degree worth?' was to closely consider the key terms within it and to examine assumptions that underpinned them. This is one useful strategy for considering how to problematise questions, but there are others.

TASK 9.2

Take a look at the following questions and think of any ways that you can problematise them in order to come up with an interesting or original argument. These are all real questions taken from university modules in various disciplines, and while some will be in areas much more familiar to you than others, it should still be possible for you to interrogate them all in some way.

1. To what extent has European integration resulted in a loss of national sovereignty?
2. 'The bad end unhappily, the good unluckily. That is what tragedy means.' Discuss this statement in relation to one of the tragedies you have studied this term.
3. What evidence would you use to argue that Britain is a racist society?

NOTES

1. This example is very similar to the one that we worked through above, in that the most obvious way to problematise the question is to focus on a *key term* in the question – in this case 'national sovereignty'. Here, a consideration of what this term means (the standard definition would normally be something like 'the ability of a country to govern itself without interference'), and whether it is ever truly possible for a country to be fully in command of every decision regarding its governance, would produce a much more interesting argument than simply arguing either that European integration had or had not resulted in a loss of national sovereignty. It could also be possible to argue that in some cases integration had led to an *increase* in sovereignty by allowing 'weaker' nations to resist pressure and aggression from other states (for example, Finland in relation to Russia). While this would not necessarily problematise the terms of the question, it would subvert the expected response.

2. There are two key elements to consider here. The first is to ask – where does the quote that the question asks you to discuss come from? This type of question (quote/discuss) is common, especially in the humanities, but also in the social sciences, and when answering such questions, it is absolutely vital that you first establish where the quote has come from. This tells you an awful lot about what you are dealing with – even if the answer is that your lecturer made up the quote. In this case, the quote is taken from *Rosencratz and Guildenstern are Dead* by Tom Stoppard, which is a play that takes two relatively minor characters in Shakespeare's *Hamlet* and tells the action from their point of view. It is a comedy, and the line quoted here is spoken by one of a group of 'players' (that is actors) within the play, who are about to perform. It is thus a self-reflexive comment on the nature of the action, rather than a serious statement, or one that should be taken at face value. It is also loaded with tragic irony, as the audience knows that Rosencratz and Guildenstern are themselves going to die before the play is over. We should thus be thinking not just about what the quote says, but how its context affects its meaning. This is a pithy quote, but not one that has been said as a serious comment on the nature of tragedy.

What you definitely do not want here is a response that just lists some bad characters that end unhappily and some good that end unluckily, even if it is then contrasted with some that break the mould. This basic, surface reading might be a place to start, but it is only that, a place to start, and would definitely not constitute an original or critical essay.

The second element that we need to consider is that this is a two-part question – i.e. (1) that in tragedies the bad end unhappily and the good unluckily, and (2) that this is what tragedy means. An answer that only focused on the first half would not get a good mark. Problematising this question would involve both a consideration of this quote and what it tells us about the nature of the play it is taken from, and also what it can tell us about the nature of tragedy, which you illustrate through your chosen text – as it is also important that while *Rosencrantz and Guildenstern* is quoted, the question asks you to discuss the quote in relation to another tragedy from the module it is taken from.

Carefully choosing your text would allow you to consider what argument you wanted to make, and the choice you made would determine exactly how you could problematise the question, and in particular that part of it looking at what tragedy actually means. While we don't have time to do that here, the key lesson we can take is that where the first example showed us how a detailed consideration of the terms of a question allowed us to build a more critical and original argument, here a detailed consideration of a source text and the specific context it provides, allows us to do the same thing. While not all questions allow you to choose the text (or topic) you are going to discuss, *you always have some ability to choose the context, or specific examples that you will use*, simply by picking the evidence and sources that you draw on. Doing so carefully will enable you to build more effective arguments.

3. The final example is interesting because the way it is written is designed to draw attention to the construction of the question itself. It would be much more usual to see this phrased as 'Is Britain a racist society?' or variations thereof, but this question is very deliberately not set up that way, which opens up intriguing possibilities. Is there a presupposition here that Britain is in fact racist? Could you argue that it wasn't? Is the question about the topic itself, or more about demonstrating an understanding of different types of evidence and how they can be used? It would certainly be interesting to think about the mirror to this – namely, what evidence would you use to argue that Britain is *not* a racist society?

Whatever approach you chose, what is certain is that you are being asked to consider the very way in which questions are asked – and answers constructed – in the subject of study (in this case sociology). You are being asked to not just provide evidence to support an argument, in other words, but to justify the choices that you are making, and comment on what evidence is valid or not valid and why, and what sorts of argument you can build based on that evidence. It is therefore likely that this question was set in this way precisely to make students aware of how arguments work in the subject they are studying, and to make them think as much about how they constructed their answer as about what that answer was.

Again, this is not an approach that you are necessarily going to come across on a regular basis, but there is still a lesson here that we can apply more generally. That is, as well as *interrogating key terms* and *making evidential choices* in order to problematise a question, we can also *interrogate the nature of the question itself* and the rules around how questions are answered and arguments constructed. This links to the discussion that we had in Chapter 7 around what your discipline means by truth, or about what makes a 'good' or a 'right' answer in your discipline. Thinking about those rules, and how they inform both the construction of questions and the arguments that respond to them, is a very fruitful way of problematising the matter at hand.

Taking another approach

The third strategy for making an argument your own involves asking the question – what would happen if we took a different approach? Our first tactic involved taking the ideas of others and considering how to critically engage with them in order to create something original, while our second involved using our own critical insight to challenge the terms on which questions are being asked and answered. The third way is a combination of the two, and covers both sides of this apparent opposition – namely using the ideas of others and having your 'own' original arguments.

There are a number of aspects to think about here, some of which we have already explored to a greater or lesser extent. You could be, for example, taking an existing argument and seeing how changing an element of that argument could change the outcome. You could be thinking like a doubter and generating counterarguments (see Chapter 5) to your own position, which you can then overcome, or arriving at a new position through considering the challenges that your alternatives have suggested. You could, to put it in Laing's terms, be interrogating your initial intentional act and seeing what effect choosing a different one would have. In each case, even just considering the alternative will allow you to critique the original position and your final answer will be stronger.

Some other ways of coming up with alternative approaches could be:

A methodological change – Exploring what difference a change in methodology might make is a useful place to start as, for example, quantitative data can sometimes give us answers that we need qualitative data to properly understand and vice versa. Equally, altering an aspect of the methods used could be interesting, whether that be examining different types of source, broadening a sample or coding data differently.

Even in subjects where 'methodology' is not quite such a clear-cut element of research or building arguments, the same applies. In literary studies, for example, focusing on a close reading of the text, as opposed to reading a text through its relationship to a historical or biographical context, could be seen as using a different 'methodology'. In this sense, if methodology does not feel like a helpful term in your discipline, it might be useful to think about example 3 from the exercise above, and consider how a question has been asked, and how asking a different type of question might lead to a different type of answer.

Considering the unconsidered – Thinking about how an under-represented or considered element/event might affect an argument can also be very useful. Often, arguments are made based on dominant positions and reflect existing power structures. Thinking about whether arguments remain valid when the marginalised or powerless (or simply ignored) are considered is both useful and important.

In Chapter 5, for example, we looked at how Neil Lyndon's (1992) argument against feminism ignored the historical context for many of the 'facts' that he cited to support his position, while in Chapter 1, when considering whether or not university was a 'good thing', we used Ngugi's (1986) account of university in British-ruled Kenya to show how university is not necessarily in and of itself the force for good that we might have assumed. In both cases, thinking about what had not been said allowed us to think differently about our argument.

The level of abstraction – The Ngugi example also highlights how a different level of abstraction can affect things. By this, I mean the degree to which we are discussing ideas in a non-specific, generalised way (in the 'abstract' or in 'theory'), or in a specific context (the 'concrete' or 'particular'). In the case of whether university is a force for good, if we think only about the 'idea' of the university, it is easy to argue that it is indeed a positive thing. However, grounding the discussion in a concrete context – whether that be colonial Kenya or something else – completely alters that.

The same shifting of levels was at play in the argument I made above about what a degree is worth. To an extent, the idea of measuring university courses by the economic benefit conferred on graduates is fine in theory or as a contextual consideration, but once we apply it to specific examples we begin to see the problems. Again, there is a contrast between the abstract and the concrete, and shifting our position along the scale between the two allows us to test how the point we situate ourselves at affects the outcome of our argument.

This approach also applies from a scientific (or perhaps just physical) perspective. For example, when considering the impact of a chemical on a living being, you could seek to explore the question on a molecular, cellular, genetic, systemic or organismic level. All would be valid but would produce potentially very different results, and shifting between the levels is a useful way of interrogating positions and coming up with new ones.

Unsettling a binary – As we discussed at the very beginning of this book, many arguments are defined in terms of a binary opposition. In that example it was the idea that the purpose of the university can be mapped onto a scale with 'money' on one end and 'knowledge' at the other – and we saw in this chapter how unsettling and showing the inadequacy of that binary can be useful in building a new argument.

Most binaries are necessary oversimplifications that are useful as a means of approaching an issue or of explaining a range of positions in response to a question. However, such explanations have a way of becoming orthodoxy – that is, they become a standard answer – and the fact that they are simplifications starts to be ignored. Making sure to interrogate any binary that you come across is a good way to ensure that you don't fall into this trap.

Taking a different theoretical approach – All arguments, as we have already established in Chapter 3, are always underpinned by a particular theoretical approach or approaches. Considering the impact of a different theoretical approach can therefore generate new possibilities and ideas.

My incorporation of Ngugi into Chapter 1, for example, could be seen as demonstrating what happens when you introduce a postcolonial perspective into an argument that was otherwise taking a very Western, liberal approach. I could just as easily have included a Marxist point of view instead that argued that universities function to maintain and reinforce the hierarchical systems of capitalist oppression, or a feminist perspective on the historical role of universities in perpetuating the patriarchy.

As this shows, taking a different theoretical approach does not have to mean refuting or abandoning the old one. We could acknowledge the legitimacy of the postcolonial, Marxist and feminist critiques of the value of the university and still argue that universities have the potential to be forces for good, even if that potential is not always realised. Indeed, we could use those very critiques to inform new possibilities for that potential.

This creation of new arguments and perspectives from the combination (or collision) of theoretical frameworks is an extension of the insight already discussed, namely that the combination or comparison of different points from existing arguments is in itself a critical and creative act.

It is also important to reiterate that there is no such thing as a position that is 'outside ideology'. There is no final, neutral position from which to arrive at an answer that is free from any theoretical consideration and therefore the ultimate 'truth'. To put it another way, it is only possible to take a different approach, and not to take no approach at all.

It is common in politics, for example, to present your opponent's views as determined by an overly rigid adherence to an ideological position while casting your own as the result of common sense or of unbiased rational thought. What this means, in other words, is that your opponent is brainwashed while you can see the real issue, or that your approach to, for example, stopping the spread of a global pandemic is based on 'the science', whereas the strategy you disagree with is determined by a belief system and an associated inability to rationally interpret the facts.

Unfortunately, this is not confined to the political sphere and the same arguments occur in an academic context. Witness, for example, how thinkers like Jordon Peterson or Steven Pinker present their own positions as being a dispassionate engagement with the truth (albeit positions that are somehow always brave and against the grain of a perceived popular opinion, despite their bestselling status), whereas those of anyone who takes issue with them are examples of poor scholarship or blind following of the herd. Not all examples are this blatant, but they are indicative of the ways in which academia is not always as objective as it claims to be.

I am not arguing here that there is no such thing as 'truth', or that there are not positions that are incorrectly adhered to because of a pre-determined belief. What I am arguing is that it is not possible to *not* be occupying an ideological position, and that 'truth' is always a value determined within a given framework, not outside of it. As Althusser says, ideology is our imaginary relationship with our real conditions of existence, and whatever those real conditions are, our relationship to them is always imaginary. This is not a bad thing or a failure, it is simply a reflection of our limitations as humans. Stephen Hawking (1988) says something similar when he argues that any theory is a model of reality that we come up with to try and understand reality, not that reality itself. This is thus not an argument that applies only to the humanities, or philosophy, or politics, it is a fundamental tenet of the scientific process – namely, that the answers you get are determined by the questions that you ask and the models you build to answer them, as much as they are any external reality that you are asking those questions about.

Why that matters for you reading this book is because of what it tells you about arguments – namely that all arguments (including your own) are underpinned by certain assumptions and choices (which I am here referring to as theoretical frameworks or ideologies). These are usually choices made in an attempt to present the most true version of reality, or perhaps more accurately to answer a particular question most effectively, and some ways will definitely be more effective than others. But that does not change the fact that it is hugely important to not allow yourself to think that your position, or the position you most agree with, is somehow free of all theoretical considerations and is in fact just 'true' and for all time. Any question is asked in a particular context based on a particular set of choices and contingent on context, and so is the answer to that question.

To come back to our overarching question for this chapter – the truth is that no argument or position is ever truly or entirely original. You are always standing on the shoulders of giants, as Isaac Newton put it. In other words, and perhaps more accurately, your own ideas are always assembled from the ideas of others, and what is new is the particular combination produced as they are filtered through the unique individual (you) and the unique context of the particular time and place in which you are asking and answering the question. This is not to remove or devalue the idea of the new – it is instead to urge you to think about the original or creative in an academic context as the product of continual questioning and iterations of rigorous research and thought, and not as the outcome of individual moments of astounding genius. It is tempting to make the history of knowledge a story of amazing individuals that made remarkable discoveries entirely on their own, but this is very seldom how it works.

To build good arguments at university then, you are required to demonstrate your own independent thought and critical insight. But this does not have to involve the creation of something entirely new. Indeed, that is arguably impossible. What you do have to produce, however, in order to get a good mark is something that is demonstrably *yours*, that shows what you have taken from elsewhere and what you and the academic context you occupy have brought to bear.

To achieve this, you can examine the assumptions that underpin the arguments and ideas in the area you are studying, problematise the questions that are being asked, consider alternative approaches, and explore whether there are unconsidered elements or areas. Most important, though, is to remember that the arguments you build are made from and always engaged with all of the other arguments that make up your discipline of study, and that all of these are always changing as new questions are asked and answered.

Summary

- In order to get a good mark at university, your arguments have to include original and independent insights and not just repeat the ideas of others.
- This does not mean that you need to come up with something completely new in every argument that you make – that would be impossible. Instead, you are simply required to always be active and critical and make your position on all the evidence that you use clear.
- You can also build original arguments or make critical points by problematising questions and considering alternative approaches.
- Examining the theoretical frameworks that underpin arguments, including your own, and exploring the impact that applying different perspectives can have is a key element of critical argumentation, as it demonstrates an understanding of not just what arguments are made about a subject, but also how and why those arguments are made.
- There is no such thing as a position or argument that does not rely on these frameworks at all, and it is vitally important to always be looking to question the assumptions of others, along with your own.
- Understanding the frameworks that are used in your own discipline and how they interact is important in becoming a successful student of that discipline, and to build good arguments you need to learn these as well as the content of those arguments.

FURTHER READING

Chapter 1: The existential-phenomenological foundations for a science of persons from Laing, R.D., *The Divided Self* (Penguin Classics, London, 2010).

Chapter 6: How to be original from Bonnett, Alistair, *How to Argue*, 2nd edn (Pearson, Harlow, 2008).

NOTE

1. These are all taken from the examples quoted in Chapter 1.

Conclusion

By now it will have become clear that this book is itself an argument – an argument about what arguments can be said to be. It takes a certain approach to that question, and hopefully also gives you everything you need to decide whether or not you agree with that approach and to develop your own.

The purpose of this book was not to give you a how-to manual, or a series of rules to follow. There are plenty of other academic skills books that do that and do that very well – I have recommended some of them throughout here. Instead, I wanted to give you ideas and sets of questions that would allow you to approach your own studies in an open, active and critical manner, and to work through various example arguments with you so that you could understand a particular approach to knowledge and to academic study – one that sees an academic discipline not as a fixed body of facts to be learned more or less well and then repeated back as needed, but rather as a collection of arguments and responses to questions about the world that is continually developing and evolving.

In this way of looking at things, studying a subject at university is not to passively accept what is presented to you but to enter into a conversation. What you are studying is therefore that conversation – a collection of shared interactions – and not a fixed set of artefacts that is the property of one person or group. This is a conversation in which you as a student participate, and the subject you study is therefore everything that is written and said about it, and not something separate or discrete, which that writing and speech is talking about.

This is important. If the primary purpose of the university, as I've argued here, is the production and transmission of knowledge, then the model of a conversation gives us a way of conceptualising what that looks like, and how it works. Vitally, it is not a transaction, where a set and defined product passes from a producer to a consumer, but a process, a constant reiteration of question and argument, argument and question, in which student and academic work together. Argument is both the outcome of this process and the process itself. It is the answer and the work of answering.

It is also all-pervasive. Every aspect of your learning as a student will engage with arguments, whether it be understanding them, challenging them or building them. Even 'facts' are not an exception, as all facts are themselves, on some level, the result of a series of choices, assumptions and claims, and cannot be discussed without recourse to arguments about what they mean. This is as true for the hardest of sciences as it is for the 'softest' of humanities subjects.

This makes everything sound quite complicated, and having said that we are not here to arrive at a set of rules. What then, can we hold on to? Despite the complexity, there are some anchor points that we can use to help us. The first of these is a broad definition of the term itself. An 'argument', for the purposes of university study, is a position taken in response to a particular question or issue, supported by propositions or reasons why that position is convincing or true. It is the word used for both that final position and for the process used to get there. That is our starting point.

The second is that arguments do not exist separately from their contexts, but are products of them. Every subject, every university, every country and every culture brings factors to bear that affect how arguments are constructed and received, and every genre uses argument in a different way within these contexts. Different theoretical approaches and ideological standpoints affect not only what arguments are made, but what arguments it is possible to make. As a student, you will need to learn how these things work in the educational context you find yourself in, but the fact that this is something that you need to learn also tells you that it is something you can practise and get better at. University study is a game that has existing roles and rules. Every different context within that may have different rules, but by thinking of it in this way, you can work out what those rules are, what your role is within that game and how you can play the game most effectively.

The same is true when it comes to understanding arguments. As we have established repeatedly throughout this book, much of what you will be expected to do as a student will involve engaging with the work of others, and understanding, deconstructing and reconstructing their arguments in order to produce your own. This is a key part of the game, and your arguments often do not have to represent the entirely new, but can rather simply be your position in an existing debate. Whether you are building or engaging with an argument, then, you should be thinking about what your purpose is, who you are writing for and what role you are playing. This will help you to understand your context.

The third thing to remember is that all arguments – and, indeed, all university disciplines – are a response to questions. Sometimes it is obvious and at other times the exact question being answered is less clear. Either way, being alive to this will help you to be an active and critical student. As you move through your studies, and your life, try to make sure that you are always thinking about why you are doing what you are doing, and what you expect and are expected to get out of it. Don't only focus on the specific task or activity in front of you but instead try to keep half an eye on the bigger picture and how the two relate. If you can manage this, then you will be much more likely to be successful.

The ability to focus on the macro as well as the micro, on the big picture as well as the individual detail, is also an important way of being able to think of alternative arguments and positions. Thinking about what questions an argument is responding to, how those questions have been constructed, and the assumptions and frameworks behind them, will give you both a better change of answering your own questions effectively, and the chance to demonstrate when it is not just the wrong argument that is being made, but the wrong question that is being asked.

In other words, writing and reading arguments, building and understanding them, are not opposites, but all part of the same process. As such, there are no 'rules' that you should be following, but rather decisions that you have to make based on a set of key questions. It is your job to work out what those questions are for your studies, but a key one to start with is always 'why?'. The key to building good arguments is understanding the context you are working in, and how the questions you are engaging with are born out of and helping to shape that context. Academic arguments, in other words, are not about the subject or separate from it – arguments *are* the subject, and learning how to argue in a particular discipline is thus like learning to speak its language. And like learning any language, it is best done by diving in.

Enjoy!

FINAL CHECKPOINT: WHY ARE YOU HERE?

In the first chapter, I asked you why you were here, reading this book, and what you wanted to get out of it. Now, I would ask you to look back on what you said then and see whether or not you have got what you wanted. I would also like you to reflect on what you have learnt from what you have read here, and to think of *at least three things* that you will implement in your future studies or life in general.

Bibliography

Adams, Bridget, *The Psychology Companion* (Palgrave Macmillan, Basingstoke, 2009).

Altbach, Philip G., 'The imperial tongue: English as the dominating academic language', *Economic and Political Weekly*, Vol. 42, No. 36 (2007), pp3608–611.

Althusser, Louis, *Lenin and Philosophy and Other Essays*, trans. Ben Brewster, (Monthly Review Press, New York, 2001).

Arshad, Rowena, 'Decolonising the curriculum – how do I get started?', *Times Higher Education Campus* (14 September 2021). Available at www.timeshighereducation.com/campus/decolonising-curriculum-how-do-i-get-started (accessed 1 March 2022, 1400GMT).

Bonnett, Alistair, *How to Argue*, 2nd edn (Pearson, Harlow, 2008).

Bradshaw, A.M., 'Physics from the inside', *Nature*, Vol. 412, No. 6843 (12 July 2001).

Bruce, Ian, 'Is "critical thinking" a useful concept in teaching EAP?', *Teaching EAP* (25 June 2020). Available at https://teachingeap.wordpress.com/2020/06/25/is-critical-thinking-a-useful-concept-in-teaching-eap-if-so-where-does-it-fit-into-our-crowded-curriculum/ (accessed 28 July 2022, 1200GMT).

Burns, Tom and Sinfield, Sandra, *Essential Study Skills: The Complete Guide to Success at University*, 4th edn (Sage, London, 2016).

Butterworth, Jon, 'Whether we find the Higgs Boson or not, particle physics is a benefit to us all', *The Guardian* (12 December 2011). Available at www.theguardian.com/commentisfree/2011/dec/12/higgs-boson-particle-physics-benefit (accessed 10 November 2022, 1700GMT).

Chatfield, Tom, *Critical Thinking* (Sage, London, 2018).

Collini, Stefan, *What Are Universities For?* (Penguin, London, 2012).

Docherty, Thomas, *Universities at War* (Sage, London, 2015).

Eagleton, Terry, *Culture* (Yale University Press, USA, 2016).

Eagleton, Terry, *Literary Theory* (University of Minnesota Press, USA, 1996).

Engel, S. Morris, *Fallacies and Pitfalls of Language* (Dover Publications, New York, 1954).

Graeber, David and Wengrow, David, *The Dawn of Everything* (Allen Lane, London, 2021).

Gray, Jasmin, 'University of Essex Feminist Society accused of discrimination over gender pay gap bake sale', *Huffington Post* (7 March 2017). Available at www.huffingtonpost.co.uk/entry/university-essex-feminist-society-discrimination-gender-pay-gap-bake-sale_uk_58be96b3e4b033be146841b2 (accessed 26/01/2022, 1032GMT).

Hawking, Stephen, *A Brief History of Time* (Bantam, London, 1988).

Hooley, Tristram and Mellors-Bourne, Robin, 'A new way to measure graduate success', *Institute of Student Employers* (3 December 2020). Available at https://insights.ise.org.uk/policy/blog-a-new-way-to-measure-graduate-success/ (accessed 28 July 2022, 1200GMT).

Houvaros, Shannon and Carter, J. Scott, 'The F word: college students' definitions of a feminist', *Sociological Forum*, Vol. 23, No. 2 (June 2008), pp234–256.

Huff, Darrell, *How to Lie With Statistics* (Penguin, London, 1991).

Hyland, Ken, 'Sympathy for the devil? In defence of EAP', *Language Teaching*, Vol. 51, No. 3 (July 2018), pp383–99.

Kahneman, Daniel, *Thinking Fast and Slow* (Penguin, London, 2012).

Khan, Andrew and Onion, Rebecca, 'Is history written about men, by men?', *Slate* (6 January 2016). Available at www.slate.com/articles/news_and_politics/history/2016/01/popular_history_why_are_so_many_history_books_about_men_by_men.html (accessed 28 February 2022, 1300GMT).

LaCapria, Kim, 'Did Trump say Republicans are the dumbest group of voters?', *Snopes.com* (16 October 2015). Available at www.snopes.com/fact-check/republicans-dumbest-group-of-voters/ (acessed 13 January 2022, 1500GMT).

Laing, R.D., *The Divided Self* (Penguin Classics, London, 2010).

Leach, Robert, *The Politics Companion* (Palgrave Macmillan, Basingstoke, 2008).

Li, Wei, 'Translanguaging as a political stance: implications for English language education', *ELT Journal*, Vol. 76, No. 2 (2022), pp172–82.

Lyndon, Neil, *No More Sex War: The Failures of Feminism* (Sinclair-Stevenson, London, 1992).

Mabbett, Ian, *Writing History Essays*, 2nd edn (Palgrave, London, 2016).

Mallard, Graham, *The Economics Companion* (Palgrave Macmillan, Basingstoke, 2012).

Mangan, Michael, *The Drama, Theatre and Performance Companion* (Palgrave Macmillan, Basingstoke, 2013).

McPeck, John E., *Critical Thinking and Education* (Routledge, London, 1981).

Metcalfe, Mike, *Reading Critically at University* (Sage, London, 2008).

Morgan, Brian and Ramanathan, Vaidehi, 'Critical literacies and language education: Global and local perspectives', *Annual Review of Applied Linguistics*, 25 (March 2005), pp151–169.

Ngugi wa Thiong'o, *Decolonising the Mind: The Politics of Language in African Literature* (East African Educational Publishers, London, 1986).

Parker, Rhiannon, Larkin, Theresa and Cockburn, Jon, 'A visual analysis of gender bias in contemporary anatomy textbooks', *Social Science and Medicine*, 180 (May 2017), pp106–13.

Pettinger, Lynne and Lyon, Dawn, 'No way to make a Living.Net: exploring the possibilities of the web for visual and sensory sociologies of work', *Sociological Research Online*, Vol. 17, No. 2 (May 2012).

Prose, Francine, *Reading Like a Writer* (Aurum Press, London, 2006).

Ramage, John D., Bean, John C. and Johnson, June, *Writing Arguments: A Rhetoric with Readings*, 5th edn (Allyn & Bacon, Boston, 2001).

Rogers, Carl, 'Communication: its blocking and its facilitation', *ETC: A Review of General Semantics*, Vol. 74, No. 1/2 (2017), pp129–35.

Rush, Dave, *Build Your Argument* (Sage, London, 2021).

Scranton, Roy, 'Raising my child in a doomed world', *New York Times* (16 July 2018). Available at www.nytimes.com/2018/07/16/opinion/climate-change-parenting.html?action=click&pgtype=Homepage&clickSource=story-heading&module=opinion-c-col-right-region®ion=opinion-c-col-right-region&WT.nav=opinion-c-col-right-region (accessed 10/11/2022 2200 GMT).

Swales, John M. and Feak, Christine B., *Academic Writing for Graduate Students: Essential Tasks and Skills*, 3rd edn (University of Michigan Press, Michigan, 2012).

Swales, John M., 'English as Tyrannosaurus Rex', *World Englishes*, Vol. 16, No. 3 (1997), pp373–82.

Thomas, Francis-Noel and Turner, Mark, *Clear and Simple as the Truth: Writing Classic Prose* (Princeton University Press, Princeton, NJ, 1994).

Toulmin, Stephen, *The Uses of Argument* (Cambridge University Press, Cambridge, 2003).

Van Emden, Joan and Becker, Lucinda, *Presentation Skills for Students*, 3rd edn (Bloomsbury, London, 2016).

Wallerstein, Immanuel, 'The French Revolution as a world historical event', *Social Research*, Vol. 56, No.1 (Spring 1989), pp33–52.

Westfall, R., 'Newton and his Biographer', in Baron, S.H. and Pletsch, C. (eds), *Introspection in Biography: The Biographer's Quest for Self-Awareness* (Routledge, New York, 1985), pp175–189.

Weston, Anthony, *A Rulebook for Arguments*, 4th edn (Hackett, Indianapolis, IN, 2009).

Williams, Raymond, *Keywords* (Croom Helm, Kent, 1976).

Index

Page numbers in *italics* refer to Figures and Tables.

Milton Keynes UK
Ingram Content Group UK Ltd.
UKHW051628021224
3319UKWH00047B/1529